COST ANALYSIS OF
PLASTIC INJECTION
MOLDS

A STEP BY STEP GUIDE TO ESTIMATE THE
FINAL COST OF PLASTIC INJECTION MOLDS

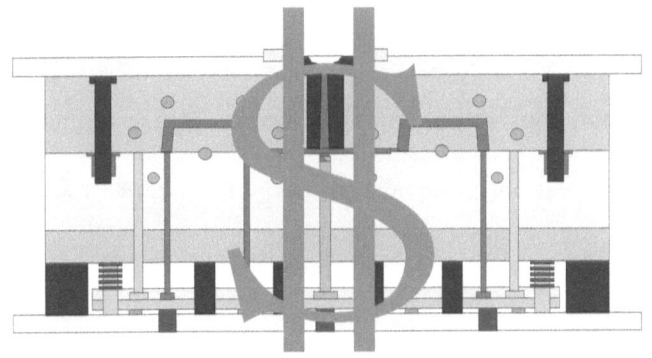

By

Carlos Sapene

Master of Science in Engineering Management

Bachelor of Science in Mechanical Engineering

2

inaccuracies, omissions, or any inconsistency herein. Any slights of people, places, or organizations are unintentional.

ISBN 978-1-4303-0295-7

First printing 2007.

ACKNOWLEDGEMENT

This book is dedicated to my parents Astrid and Carlos Sapene for supporting and encouraging me over the years. And to my lovely wife Tabatha Ann Sapene without her patience, love and support this accomplishment would not have been possible.

PREFACE

The ultimate goal of the cost analysis of plastic injection molds is to specify a plastic injection mold capable of making a cost effective product during the entire life of the project and to determine its final cost.

Achieving a cost effective mold is a very complex activity that could very well make the difference between generating a profit or assuming a significant loss in any particular job.

Most of the literature available in the industry for cost analysis of injection molds is developed primarily for specialists with a scientific background. Therefore, the existing literature presents the cost analysis of injection molds in in-depth engineering calculations to justify their formulas. This approach creates more conflict than benefit for non-technical personnel in the business community.

Performing a cost analysis for plastic injection molds varies by company and on a case-by-case basis. In many cases the people leading this activity do not have the technical background to understand the application of this process and its possible costly impact on the organization. This could very well result in a very vague tooling specification, which causes a "guesstimate" on the part of the toolmaker in order to submit a quote.

Generally, the business is awarded to the lowest bidder. There are so many assumptions and misdirection on the tooling specification that none of the quotes are alike and the final result is a poor quality product that does not meet the company's expectations. In conclusion, this makes the cost analysis of plastic injection tooling a very complex activity that must be carefully evaluated to generate a profitable product, which will allow a company to succeed in this highly competitive industry.

In summary, this book can be used as a step-by-step guide to understand and assess a cost efficient selection of plastic injection mold(s) for any personnel in the business and/or technical community.

8

TABLE OF CONTENTS

12

TABLES

INTRODUCTION

This book presents a step-by-step methodology to effectively estimate the final cost of plastic injection molds in the industry using basic engineering calculations, financial management analysis, strategic planning, and project management tools in order to assess the financial considerations through technical specifications.

This analysis takes in consideration all the elements and the necessary activities that directly or indirectly affect the final cost of the plastic injection mold, from part design approval up to mass production. In addition, this analysis visits each one of the cross functional activities inside of an organization to develop a tooling specification that meets the company's requirements for a particular project. This tooling specification is used to estimate the final cost for a plastic injection tool, breaking down each element in the process of making the tool and taking in consideration any required additional activity during the development of the project.

The different approaches from toolmakers to quote a specific tool, the advances in technology, and the lack of required knowledge of the customers for laying out a tooling specification sheet have created major problems to accurately estimate the final cost of plastic injection tools in the industry. Very often these tooling specification sheets do not translate the company's needs into words that the toolmakers can understand in order to make the mold, thus leading to a project failure.

This book develops a step-by-step methodology to complete a thorough evaluation for a plastic injection tool in order to produce a plastic product for any specific industry, interrelating each step of the process.

The following cost analysis methodology allows for the establishment of effective allocation of resources and scheduling of activities to achieve a plastic product capable of meeting the customer's expectations within an estimated budget.

CHAPTER 1

OVERVIEW

1.1- Scope:

This book establishes a methodology to accurately estimate the final cost of plastic injection molds. The formulas used throughout this book are taken from previous literature, and will be reference for the reader's further use. It is not the author intent to develop an in depth mathematical demonstration of these formulas. This book will support the reasoning for using these formulas and will explain the relevance of these calculations through this process to arrive at a successful cost analysis of plastic injection molds. Specifications use throughout this book follows the industry standards and SPI (Society of Plastic Industry) standards. This work assumes that the parts in this study can be made throughout the plastic injection molding process. Begins with a request to make a plastic injection mold for an approved plastic part, and it concludes by mentioning some financial considerations for the building of a mold after a complete cost analysis study is made.

1.2- Influential factors in the cost analysis of the plastic injection molds:

Laying out plastic injection tooling requirements that meet the needs of a particular job is a cross functional activity that requires the involvement of many different departments of the organization. The common critical factors to consider previous to laying out the tooling specifications are the sales forecast, injection machine specifications, part design, plastic material specifications, technological requirements, quality specifications, fitting to matting parts, tooling budget, and method of payment. The analysis and understanding of this preliminary information defines the general guidelines for tooling layout specifications.

This information must be translated in a way that the toolmaker is able to understand. This translation represents a tooling specification sheet that meets the company's requirements for a specific job. For example, the tooling forecast gives you the estimated lifetime of the project and the estimated production requirements. During the translation for the toolmaker the lifetime of the product and the estimated production requirements will be equivalent to the required lifetime and usage of the tool. This information is one of the critical factors in defining the material to be used for building this specific tool. In addition, the estimated production requirement is one of the critical factors used to select the number of cavities on the tool.

Using new available technology in plastic simulation, tooling design and manufacturing, and project management among others, will give to the final tooling specification the required technological advantage to stay ahead of the competition. The tooling specification must show to the toolmaker the critical requirements for a specific job.

A complete tooling specification will be used to estimate components of the plastic injection tool and predict the amount of work that has to be done to finish this tool. Knowing the required components of the tool and being able to predict the amount of work will allow one to get an accurate estimated cost of the tool and the time involved to complete this job.

The most common additional factors affecting the final price of the mold are tooling transportations, tooling trials, travel expenses, and dimensional evaluations.

The final estimated cost of the mold is compared against previous jobs with similar requirements. These analyses will contribute to establishing the target price and the required time to build a specific tool. This target price and timing are used to evaluate the tooling quotes for this job.

The selection of the possible toolmakers to quote for building a tool is a critical decision that requires a capability evaluation of the toolmaker. The possibility of finding a place with the ability to make a lower cost tool meeting the delivery date is always there. Globalization and the advances in communication such as the Internet in the last decade have allowed companies to compete virtually from any where around the world. When an overseas toolmaker submits a quote for a project, the tooling cost analysis should include risk factors to fairly evaluate the advantages and disadvantages of this option. Sociopolitical stability, economic incentives, language barriers, and business relationships between both countries and organizations are some of the many factors to consider when building molds overseas.

1.3- The economic significance of plastic injection molds:

The ultimate goal of cost analysis of plastic injection tools is to specify a plastic injection mold capable of making a cost effective product during the entire life of the project and determine its final cost.

The evolution of the tooling and the injection molding market goes hand in hand. Plastic products are replacing glass, wood, and aluminum among other materials in the industry. The market place is always looking for materials to reduce cost, keeping or exceeding customer expectations. Plastic products are meeting these needs in many different industries.

Many industries have taken advantage of this injection molding process by meeting functional requirements and in addition, enchanting customers with creative new part designs in very competitive markets. An injection-molding machine is standard equipment that can be used for many different molds. However, each mold has to be made for particular designed part(s). A plastic injection tool is used to shape the parts through the injection molding process.

In general, when a new plastic part is designed for a plastic injection molding process the building of a new tool to make this plastic part is required. To make a particular plastic part, the injection tool can be built in many different ways.

Achieving a cost effective tool is a very complex activity that could very well make the difference between generating a good profit or having a big loss in any particular job.

1.4- Trend of the injection mold industry:

Companies struggle with the dilemma of cost, efficiency, quality, reliability, and timing among many other issues entering into the market. These factors make injection-tooling considerations a critical facet in this analysis that can make the difference between generating profit or having loses during the life of the project.

The demand for plastic parts and the furious competition force industries to develop new products in very short periods of time, and more often than ever. Plastic parts are very popular in many different industries. These make the plastic injection tooling industry a multimillion-dollar business in the USA and around the world.

The tooling industry has evolved so fast that twenty years ago toolmakers could not possibly imagine the dimensional tolerances, processes and speeds used today to make a tool. This industry continues to make many technological advances in a very short period of time. Organizations are realizing the advantage of the changes in this industry transferring this cost savings to the final product in order to increase sales and maximize shareholder wealth.

1.5- Structural Components of a Mold:

The industry has standardized many of the required elements for building a mold. This standardization allows mold makers to order many of the components of the mold by catalog. This allows mold makers to pay more attention on the manufacturing of the critical elements such as the cavity and the core blocks. This ordering of standard parts by a catalog relieves repetitive work, reducing the lead-time and in some cases giving additional cost saving to the final cost of the mold.

Knowing the necessary elements for building an injection mold is critical to identifying the component(s) that can be ordered by catalog and the component(s) that have to be manufactured by the toolmaker.

The following list shows the primary components for a plastic injection mold. Many of these components can be purchased from sub-suppliers. The toolmaker can choose to manufacture some of these components. However, it is very important to stress the need for keeping standard dimensions on them in order to simplify the replacement of parts during mass production.

1-Front clamping plate.

2- Front cavity plate.

3- Support core plate.

4-Support plate (Back up plate).

5- Spacer block.

6- Rear clamping plate.

7- Ejector plate.

8- Ejector retainer plate.

9- Support pillar.

10- Stop pin.

11- Leader pin.

12- Leader bushing.

13- Locating ring.

14- Sprue bushing.

15- Spring for ejector system.

16- Rod for ejector system.

17- Parting line locks.

18- Water connectors.

19- Eye bolts.

20- Sprue puller.

21- Ejector pin.

22- Sleeve pin.

23- Core pin.

24-Return pin.

25- Lifter action.

26- Slide action.

Standard components of an injection mold

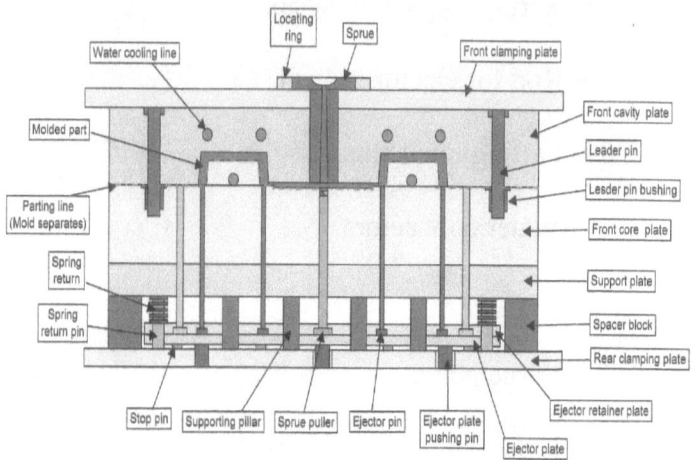

The materials for building a plastic injection mold and its components are very diverse. The understanding of the mold requirements allows one to make a cost effective choice during this selection.

1.6- Methodology:

The forecast analysis determines the company output for a specific product that defines the capacity planning and the quality requirements for a particular project. The forecast analysis lays out the company's expectations for a final plastic injection molded part.

These company expectations establish the part's specifications that are used to create a mold specification sheet. This mold specification sheet lays out all of the requirements for building a tool capable of making parts that meet all of the company's expectations.

This activity requires a good understanding of the plastic injection mold structure and its effects on the final plastic parts.

The right selection of mold components, in addition to, choosing the best alternatives among all the different available options for building the mold are the most critical activities at this time. The plastic injection mold must be able to make plastic parts within specification for a competitive price during the life of the project.

To successfully estimate the final cost for a plastic injection mold, it is necessary to determine all the elements involved during the mold building process. It is very important during this cost estimation process to establish which elements are going to be manufactured by the mold maker and which elements are going to be purchased as standard items from a sub-supplier catalog. The final cost of the mold represents the sum of the cost of every element of the mold, plus the cost for any additional activity involved during the mold

building process.

CHAPTER 2

CYCLE TIME STUDY

Overview of the analysis

The allocation of resources and the cost of producing plastic parts are directly affected for the cycle time of the process. This is the required time for the injection-molding machine to complete one full cycle through out the injection molding process in order to produce plastic parts. The number of cavities of a plastic injection mold establishes the amount of parts that the machine is capable of making per each cycle time. The cost of a plastic injection mold increases proportionally to its number of cavities.

The cycle time and the capacity requirements are necessary pieces of information for the selection of the optimum number of cavities for a plastic injection mold. A plastic injection mold with too many cavities is more expensive to produce and needs a bigger machine to make the plastic parts. On the other hand, a mold with not enough cavities is not capable of meeting production requirement(s) thus, shutting down the production line.

To estimate the cycle time during the injection molding process it is necessary to understand the elements involved in this activity. In order to simplify the analysis and highlight the impact of each one of these elements on the overall cycle time, they have been grouped in four main categories: Plastic material, part design, injection molding machine, and injection mold. These groups are studied separately and integrated later on during the final decision making process.

2.1- Plastic material and part design: The properties of the plastic material and the part design define the material flow and the maximum heat exchange rate allowed from the part to the mold. The material flow affects the injection time and the heat change rate affects the cooling time.

There are complete fields of study dedicated to the analysis of plastic materials. It is common practice to rely on the material suppliers for material specifications.

2.2- Injection molding machine: The cycle time for the injection molding machine is separated into three principal stages: injection time, cooling time, and ejection time.

2.2.1- Injection time is estimated using the material flow rate during the injection stage and the volume of the cavities required to be filled inside the mold. Normally, this stage takes less than 10% of the total cycle time.

2.2.2- Cooling time is affected for the heat exchange rate of the plastic material and the thickness of the part. The estimated temperature of the part during the injection stage (beginning of the cooling time) and the expected temperature of the solidified part at the ejection stage (ending of the cooling time) establish the amount of necessary heat to be removed from the part during the cooling stage.

The exchange rate for a specific plastic material, the gradient of temperature to be removed from the plastic part, and the thickness of the part allows one to estimate the cooling time for this process. The cooling time may consume up to 80% of the total cycle time. An accurate estimation of the cooling time significantly increases the success for the estimation of the cycle time.

2.2.3- Ejection time is the required time to release the solidified part from the mold. The injection machine opens separating the core form the cavity of the mold. Then, the injection molding machine actuates the ejection system to remove the part(s) from the mold. The solidified plastic part(s) is released from the mold; then the machine closes the mold to continue with the next cycle.

The required time for this stage depends of the ejector system used and post operation processes. This can range between five percent (5%) and 25% of the cycle time and it varies on a case-by-case basis.

2.3- Injection mold: The selection of the materials for building an injection mold and the mold designs have a direct effect on the quality of the final plastic part and its cycle time.

The mold has to be capable of removing as much heat from the part as it can transfer to the mold. If the injection mold has a poor cooling system, the heat exchange rate between the plastic part and the mold could be a limitation for the cooling time.

2.4- Estimation of the total cycle time.

The estimation of the cycle time for the injection molding process requires a thorough understanding of each one of the stages discussed previously.

2.4.1- Estimation of injection time (Ti):

The injection time is the time required for the injection molding machine to fill the cavity(s) of the mold with melted plastic. To simplify this analysis, the following preset values can be used for cost estimation purposes. An estimated filling time of four (4) seconds per parts molded in an injection molding machine smaller than 500 tons and, an estimated filling time of seven (7) seconds per parts molded on 500-ton injection machines or bigger. If the injection rate of the injection molding machine is known, the time of injection would be equal to the volume of the injection chamber (volume of the part and runner – mm^3) divided by the injection rate (machine rate - mm^3/sec).

2.4.2- Estimation of cooling time (Tc):

The heat exchange between the plastic and the coolant takes place through thermal conduction in the mold. Because moldings are primarily of a two-dimensional nature and heat is only removed in one direction (the direction of the thickness of the part), a one-dimensional computation is sufficient. Considering that the mold is capable of removing as much heat as necessary from the part. The critical value for this calculation is the time required to remove the heat from the part. The equation to estimate the cooling time based on the previous analysis is as follows:

$$Tc = [S^2 / (A * \text{Л})] / \{Ln [(8/ \text{Л}^2) * ((Tm-Tw)/(Te-Tw))]\}$$

S= Wall Thickness (mm)

Tm= Melt Temperature (°F or °C)

Te= Temperature of the part after molding (°F or °C)

Tw= Cavity wall temperature (°F or °C)

A= Thermal diffusivity (mm^2/Sec)

Л= 3.1416

The wall thickness of the part is established during the part's design, and the material specifications can be obtained from the material supplier. Some material suppliers do not show the thermal diffusivity of the plastics. Therefore, the equation to obtain the thermal diffusivity of the plastic material is as follows:

A= k / (p* Cp)

A= Thermal diffusivity (SI units: m^2/sec).

K= Thermal conductivity (SI units: watts per meter-kelvin, W/(m*°K)))

Cp= Specific heat capacity (SI units: joule per kilogram*kelvin, J/(Kg*°K))

P= Density (SI units: kilogram per cubic meter, Kg/m^3)

P*Cp= Volumetric heat capacity (SI units: Joules per cubic-meter-kelvin, J/(m^3*°K))

2.4.3- Estimation of ejection time: The major elements to take in consideration during the ejection process are the opening of the mold, the ejection of the part from the mold, and the removal of the part from the machine.

2.4.3.1- Opening of the mold: The injection-molding machine separates the core side of the mold from the cavity side of the mold to a distance with enough clearance to eject the part from the mold, thus removing it from the machine. The distance to take in consideration for the opening of the mold includes the stroke of the ejector system in order to release the part from mold, plus the space in between both sides of the mold to access and remove the part from the machine.

The formula to estimate the distance for mold opening is the following:

$Dm = Ds + Dr$

Dm = Mold opening (mm or in).

Ds = Stroke of the ejector system (mm or in).

Dr = Access to remove the part (mm or in).

The time for mold opening is estimated through the equation of linear motion uniformly accelerated from the rest:

$$Tm = \sqrt{(2*Dm / Am)}$$

Tm = Time for mold opening (Sec).

Am = Acceleration during mold opening (mm/Sec^2).

It is possible to estimate the time required to eject the part from the mold.

$$Ts = \sqrt{(2*Ds / As)}$$

Ds = Stroke of the ejector system (mm or in).

Ts = Time to complete the ejection of the part (Sec).

Am = Acceleration of the ejection system during opening (mm/Sec^2).

The minimum stroke to release the part must exceed any undercut; otherwise the part will be unable to be removed from the mold. For estimation purposes, a conservative value of 15%

over the maximum depth of the part parallel to the

direction of the ejection should be used.

Ds = 1.15 * Dp

Dp = Depth of the part (mm or in).

The three most widely used techniques for

removing plastic parts from the injection-molding

machine are removal of parts by hand, by robot,

and by free fall. Removing parts by robot reduces

the labor cost and increases the repeatability

during the injection molding cycle. Removing

plastic parts by robot has increased rapidly in the

injection molding industry becoming the norm in

many profitable injection-molding companies.

This study considers the removal of the plastic

parts by robot. Similarly, any other technique can

be analyzed following the same concept presented

below.

Part removal by robot: during the cooling time the robot returns to zero position normally located on top of the mold at the injection molding machine. After the mold opens and the part is ejected from the mold, the robot moves down in between both sides of the mold until it is parallel to the plastic part. Then, the robot moves forward horizontally to grip the part. It grips the plastic part and pulls backward to clear the plastic part from the ejector system in a horizontal motion. Next, the robot with the part moves up in a vertical motion to free the part from the machine and places it in a specified location. As soon as the part is off the machine, the injection machine receives a signal to close the mold thus starting a new cycle. Any additional operations required for the robot are not included in the cycle time for the injection molding process.

The robot uses horizontal and vertical motions during the part removal process. The time for part removal by robot can be estimated through use of the linear motion uniformly accelerated formula showed previously.

The next step is to determine the horizontal and vertical distances required for travel by the robot. On the vertical direction the robot must clear the part from the machine. To estimate the vertical distance, it is assumed that the part is located on the center of the mold for the vertical direction. Under this assumption the distance traveled by the robot is the top half of the machine plate plus the bottom half of the part, and any additional distance to clear the part.

Vertical travel (Vt):

Vt (mm or in) = ½ Mp + ½ Pv + Cs

Mp = Machine plate (mm or in).

Pv = Part dimension on vertical distance (mm or in).

Cs = Clearance space (mm or in).

For the horizontal travel the robot moves forward to reach the part. Next, the robot grips the part and moves backward to free the part from the ejection system. To estimate the horizontal distance required for travel, it is assumed that the part is free from any undercut at the ejector system when the part travels the depth of itself away from the ejector system plus an additional safety distance to clear the part.

Horizontal travel (Ht):

Ht = Dp + Cs

Dp = Depth of the part (mm or in).

Cs = Clearance space (mm or in).

The required distance travels for the robot in order to remove the plastic part from the machine is (2 * Vt) plus (2 * Ht) and the time used during part removal is estimated as follows:

$$Tv = \sqrt{(2\ Vt\ /\ Avt)}$$

$$Th = \sqrt{(2\ Vht\ /\ Aht)}$$

$$Tr = 2 * Tv + 2 * Th$$

Tr = Time required to remove the part from the machine (sec).

Tv = Time for vertical traveling (sec).

Th = Time for horizontal traveling (sec).

Avt = Acceleration on vertical direction (mm/sec^2).

Aht = Acceleration on horizontal direction (mm/sec^2).

2.4.3.2- The total ejection time: The total ejection time is calculated as follows:

Ejection time = open mold + ejected part + removed part + closed mold

$Te = Tm + Ts + Tr + Tm$

$Te = Ts + Tr + 2Tm$

Te = Ejection time (sec).

Tm = Time for mold opening (sec).

Ts = Time to complete the ejection of the part (sec).

Tr = Time required to remove the part from the machine (sec).

2.5- Total cycle time: The total cycle time for the injection molding process is the sum of the injection time, the cooling time, and the ejection time.

Total cycle time = Injection time + cooling time + ejection time.

Total cycle time (TCT) = Ti + Tc + Te

Today, flow analysis simulation software for plastic injection molded parts is capable of estimating the cycle time, and it can be purchased from many different suppliers. The person running the simulation software must have the necessary skills to predict the real condition during the injection molding process in order to obtain accurate results from this analysis.

In any case, a full understanding of the injection molding process is necessary to effectively complete this activity.

CHAPTER 3

SELECTION OF THE INJECTION MOLDING MACHINE

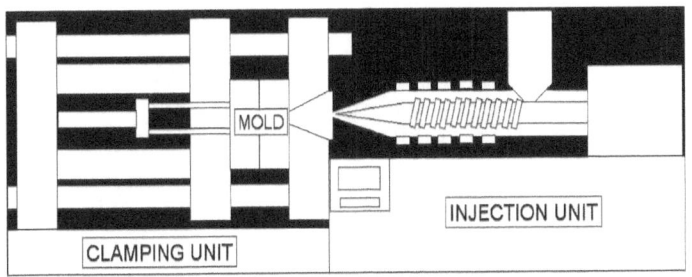

The right selection of injection molding machines is a very important activity that should not be underestimated. Any variation on the size of the machine will affect the final quality and/or cost of the plastic part. The injection molding machines must have sufficient clamping force to keep together both sides of the mold during the injection, and the injection mold must fit between the tie bars of the machine. The elements to take in consideration during the selection of an injection molding machine for a specific injection mold are the clamping force, the distance between tie bars, and the shot size.

3.1- Clamping Force: The clamping force of the injection-molding machine must exceed the force created inside the cavity of the mold during the injection stage. The function of the clamping system is to keep both the cavity side and the core side of the mold together during the injection stage in order to fill the cavity of the mold with melted plastic. The melted plastic forced inside the cavity of the mold is the product of the cavity pressure through the total projected area of the part. The projected area of the part is the shadow falling on

a plane surface parallel to the parting line.

Clamping force (Ton)= Projected area (cm^2) x

Plastic pressure cavity (Kg/cm^2).

$Cf = Pa * Pp$

Cf = Clamping force (Tonnage)

Pa = Projected area (cm^2)

Pp = Plastic pressure cavity (Kg/cm^2)

A cavity pressure of 300 Kg/cm^2 is a

reference value used to calculate the required

clamping force to keep the mold closed during the

injection stage.

3.2- Distance between tie bars: The

injection mold must fit in between the tie bars of

the injection-molding machine. The dimensions of

the clamping plate define the overall size of the

mold, and these dimensions must be smaller than

the distance between the tie bars of the injection-molding machine. The size of the clamping plate of the mold is determined by: the arrangement and quantity of part(s) inside the cavity, the runner system, the ejection system, the type of actions, and/or any other element necessary used in the mold in order to make the plastic part. The final dimensions of the clamping plate keep the same length (Y) of the cavity block. An additional section is required on the width (X) of the mold to allow the necessary space for clamping on the sides of the mold against the injection-molding machine.

The dimensions of the clamping plate define the maximum external size of the injection mold.

3.3- Shot size: This establishes the amount of raw material necessary for each molding cycle.

The shot size is the amount of melted plastic required to fill the cavity(s) of the mold (including runners) during one injection molding cycle.

The required shot size for a particular mold must be smaller than the amount of raw material that the specified injection molding machine is capable of loading at once in the barrel. On the other hand, a small shot size in relation to the amount of raw material loaded in the barrel of the machine increases the risk of degrading the melted material due to the number of cycles required to empty the barrel.

The degradation of the melted material in the barrel of the injection molding machine could be affected by many different factors including the temperature of the barrel, melt point of the material, and cycle time, among others. For estimation purposes, the amount of raw material loaded in the barrel of the injection-molding machine should be between 1.3 and 4.0 times the shot size.

Ss (gr) = N (unit) x P (gr/unit) + R (gr/unit).

Ss = Shot size (gr)

N = Number of cavities (units)

P = Part weight (gr/unit)

R = Runner weight (gr/unit)

Bc (gr) = Ms (mm) x Ac (mm^2) x Dm (gr/mm^3).

Bc = Barrel capacity (gr)

Ms = Maximum stroke of the screw (mm).

Ac = Effective cross section area of the screw (mm^2)

Dm = Density of the raw material (gr/mm^3).

1.3 Shot size < Barrel capacity < 4.0 Shot size.

CHAPTER 4

COST ANALYSIS OF PLASTIC INJECTION MOLDED PARTS

The cost of the injection mold is a big part of the initial investment of the project. The payoff of this investment must be considered on the final cost of the injection molded part(s). It should be calculated as a portion of the plastic part during the amortization time period for a particular mold. The part cost analysis must include all the required expenses for making a final plastic part plus the payoff of the injection mold. The part cost analysis takes in consideration the mold cost (amortization), Material cost, machine cost, and secondary processes cost.

Part cost ($/unit) = Material cost + Machine cost + Manpower cost + Secondary cost + Mold amortization cost.

4.1- Material cost: This is estimated using the weight of the part (Kg) by the material cost ($/Kg).

Material cost ($/unit) = Weight of the part (Kg/unit) * Material cost ($/kg).

4.2- Machine cost: It considers the machine operation cost ($/hr) by the cycle time of the mold (sec) divided by the number of cavities on the tool.

Machine cost ($/unit) = Machine cost ($/hr) * Cycle time (sec) * (1hr/3600sec) / number of cavities (unit)

4.3- Manpower cost: It includes the machine operator cost plus any additional manpower required during the molding process. Sometimes one person takes care of more than one injection-molding machine at the same time. In this case, a percentage must indicate the time dedicated from this person to service the machine making the parts in this study.

Manpower ($/unit) = (%) * Operator ($/hr) + (%) * Additional manpower ($/hr)) * Cycle time (sec) * (1hr/3600sec) / Number of cavities (unit)

4.4- Secondary processes cost: It includes all the operational costs for making and supplying the parts in addition to the material cost and the machine cost. The most common elements to take in consideration are the assembly cost, packaging cost, set up cost, manpower cost, and shipping cost.

Secondary cost ($/unit) = Assembly cost ($/unit) + Packaging cost ($/unit) + Shipping cost ($/unit) + Set up cost ($/unit) + Manpower cost ($/unit)

4.5- Mold amortization cost: This value represents the payoff of the injection mold for a particular project during a specified period of time. This may or may not be the life of the project. This mold cost is included in the part cost during the payoff period.

Mold amortization cost = [Mold cost ($) / Number of cavities (unit)] / [Payoff period (years) * Capacity (unit/year)]

64

CHAPTER 5

BUBGETING FOR A PLASITIC INJECTION MOLD

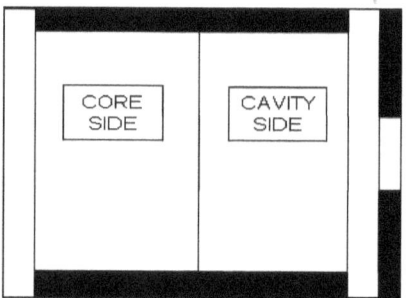

The cost estimation for a plastic injection mold establishes the budget and the necessary resources for building a plastic injection mold. In addition, it is used to determine the final part cost for the plastic product. Creating a budget for the building of a plastic injection mold requires a full understanding of the plastic injection mold design and manufacturing, and each one of its components.

A mold specification sheet lays out the general requirements for mold building. It is very important to know what the specification requirements are for each mold. An injection mold cost estimation analysis should break down the components of the mold into single elements. It is very important to identify the components that can be ordered by catalog and the components required to be manufactured by the toolmaker. For example, mold bases can be ordered by catalog from a sub-supplier. However, the cavity must be manufactured by the toolmaker to create the expected shape of the final product for each mold, which can be approached in many different ways. The mold specification sheet lays out the customer expectations for a particular mold. This allows the mold maker to establish preset requirements throughout the manufacturing process in order to achieve this expectation.

For cost estimation purposes it is critical to identify the required components for a particular mold. At this point the number of cavities and the production requirements are known values.

To simplify this analysis the plastic injection mold has been divided into the following main groups: part analysis, structure of the mold, ejection system, injection system, and, cavity and core.

5.1- Part analysis: The part shape dictates the elements required for the construction of the mold in order to make the plastic part. Stylus designers establish the parting line of the mold. This parting line is the matting line between the cavity and the core of the mold. Any undercut on the part normal to the direction of the ejection movement requires an additional action to release the part from the mold (for example: lifters or slides). These actions have direct effect on the final cost of the mold.

5.2- Structure of the mold: The injection mold requires a basic structure to hold all the elements in place. These elements such as mold bases, rails, support pillars, among others can be ordered from a sub-supplier catalog. This type of components can be specified and priced without major complications.

5.3- Ejection system: The ejection system considers all the necessary elements to eject the part(s) from the mold. The shape of the part plays a critical role in this analysis. Slides, lifters, and other components involved in the ejection of the part are considered at this time.

5.4- Injection system: The injection system includes all the necessary elements to transport the melted plastic from the injection machine through the mold to the part (runners, gates, sprue bushing, etc).

5.5- Cavity and core: The cavity and core blocks are pieces of steel manufactured in such a way, that when they are closed against each other, they give the final shape to the melted plastic. The cavity and the core layout take in consideration the parting line shut off, the projected area of the molded part(s), and any action that is in contact with the part. The cavity and core are manufactured by the toolmaker. The mold specification sheet must show the requirement for manufacturing of the cavity and the core of the mold.

CHAPTER 6

STANDARD
COMPONENTS

To estimate the price for the manufacturing of a plastic injection mold it is necessary to predict the cost for each individual element involved in making the mold. To simplify this cost analysis activity the elements of the injection mold are separated into two main groups: standard components (which can be purchased from a sub supplier catalog), and the parts manufactured by the toolmaker.

Standard components: To determine the final cost for these components it is necessary to estimate the material, the critical dimensions of each component and its required quantities in order to proceed with the selection from a supplier catalog. The sum of all these costs reflects the final cost for standard components. This value is later added to the final equation for estimating the total cost of the plastic injection mold. The considerations for the selection of each standard component are shown below.

6.1- Front cavity plate: The front cavity plate is the part of the mold on the stationary side of the machine in direct contact with the plastic material. The thickness of the front cavity plate must consider space for cooling lines and enough steel material to preserve the structural integrity of the mold. These dimensions can be adjusted to select a standard plate from a sub-supplier catalog.

6.2- Front core plate: The front core plate is the part of the mold on the movable side of the machine in direct contact with the plastic material. The thickness of the rear plate takes in consideration the space for cooling lines, and additional actions like lifters and slides, among others. These dimensions could be adjusted to select a standard plate from a sub supplier catalog.

Estimating the dimensions of the front cavity plate and front core plate: The front cavity plate and the front core plate shape and cool the plastic part(s). The external dimensions of the front cavity plate and the front core plate are very similar and they can be calculated in a similar way. The dimensional estimation of the front cavity plate and front core plate is presented below:

The cavity layout depends on the number of parts machined on the cavity plate. The sum of the projected areas of the total amount of part(s) plus an additional one (1) inch in between each part define the dimensional estimation for the cavity layout of the mold.

The estimation of the external dimensions of the front cavity plate and the front core plate are the following: the width of the mold (X) defines the horizontal dimension of the mold, the length of the mold (Y) defines the vertical dimension of the mold, and the depth of the mold (Z) defines the thickness of the mold.

X = Horizontal, Y = Length, Z = Depth

The width of the mold (X) takes in consideration the cavity plate and core plate layout plus the necessary space to locate the leader pins, return pins, and the spacer blocks of the injection mold. These elements are located on the outside of the cavity and core plate layout but they must be included inside of the cavity and core plates final dimensions. Estimating the dimensions of these elements depend on the size of the cavity and the core layout. Bigger molds require bigger components. This situation increases the overall dimensions on the mold. If there is not enough information to determine these values, a rough estimation of four (4) inches for each side of the mold is a conservative number that can be used during in this case. The result of this approach is an additional 8 inches on the total width of the front cavity plate from the actual dimensions of the cavity layout.

X (in or mm)= (Mold layout on X direction) + (8 inches)

X (in or mm)= ((Projected area on X direction in inches + 1 inch) * (# of cavities on X direction)) + (8 inches)

The length of the mold (Y) takes in consideration the cavity and core layout plus an additional space for the return pins, and leader pins. Estimating the dimensions of these elements depends on the size of the cavity and the core layout. Following similar analysis than the shown for X direction.

Y (in or mm)= (Mold layout on Y direction) + (8 inches)

Y (in or mm)= ((Projected area on Y direction in inches + 1 inch) * (# of cavities on Y direction)) + (8 inches)

The depth of the mold (Z) is estimated by multiplying the depth of the part in the direction of the mold opening (Z-direction) by 2.25. This value estimates the thickness required to keep the structural strength of the mold, including the water lines and the action elements.

Z (in or mm)= (depth of the part) * 2.25

These X, Y, and Z dimensions are checked against standards available for mold bases, this selection must satisfy the estimated dimension (X,Y,Z).

The required clamping force and the mold external dimensions are used to establish the size of the injection machine capable of making parts from a specific injection mold. The mold must fit between the tie bars of the injection machine moreover; the injection machine must have enough clamping force to keep the mold closed during molding.

6.3- Locating ring: The locating ring is placed in the center of the injection mold on the back of the cavity clamping plate. It is used to center the mold on the machine plate. The machine plate uses standard locating ring holes to center the molds. The mold must have a locating ring capable of fitting the dimensions specified on the machine specification sheet. This locating ring is a standard component, which can be purchased from a sub supplier catalog.

6.4- Sprue bushing: The sprue bushing is a standard component that guides the melted plastic from the injection machine into the mold.

The dimensions to consider during the selection of the sprue bushing are the machine nozzle tip orifice, machine nozzle tip radius, and sprue bushing length. In some cases a heated sprue bushing is required to increase the flow of the plastic material passing thought out this channel.

6.4.1- Nozzle tip orifice: The nozzle tip orifice is the diameter on the tip of the nozzle where the melted plastic is transferred to the sprue bushing.

6.4.2- Nozzle tip radius: The nozzle tip radius is used to identify the outside radius of the nozzle tip that lands on the outside of the radius of the sprue bushing. This radius must be sealed off to avoid any leaking during the injection of melted plastic coming from the machine into the mold.

6.4.3- Length of the sprue bushing: The length of the sprue bushing is determined for the distance between the cavity clamping plate and the runner of the mold. The sprue bushing transports the melted plastic from the machine to the runner of the mold.

6.5- Front clamping plate and rear clamping plate: There are many different techniques for clamping an injection mold to the injection-molding machine. These clamping systems range from sophisticated automated clamping to direct manual bolt clamping. If the clamping system is not specified during the cost estimation stage, the following considerations are used to estimate the final dimensions of the front clamping plate and rear clamping plate.

The dimensions of the front and rear clamping plates of the mold must exceed the dimensions of the cavity and core plates of the mold, in order to allow enough space to clamp the mold to the injection machine. It is recommended to add two (2) inches per each side of the front cavity and core plates of mold for this purpose. In other words, if the clamping system is not specified, the width of the front and rear clamping plates should be at least four (4) inches bigger than the cavity and the core plate (X- direction). The length of the front clamping plate should match the length of the cavity plate (Y-direction). Also, it is suggested to adjust the dimensions of the clamping plate to fit the machine and supplier standards. This adjustment must meet the technical requirements of the injection machine, and the injection mold.

The thickness of the clamping plate, and the steel type may vary case-by-case and they must be defined in order to estimate the final cost. The common thickness for clamping plates range from one half (1/2") inch to two (2) inches depending on the material and the weight of the mold.

6.6- Support plate (Back-up plate): The back-up plate is located behind the front core plate. It acts as the support plate, guiding the ejector pins, lifters, and other actions of the mold. In addition, this support plate holds the pillar bases in order to increase the strength of the mold. The dimensions of this support plate are the same dimensions of the front core plate, however, its thickness can be significantly smaller that the front core plate thickness. For cost estimation the thickness of the support plate is 50% smaller than the thickness of the front core plate.

6.7- Spacer blocks (rails): The spacer blocks are located between the back-up plate and the rear clamping plate. The spacer block creates the housing for the ejector system. One spacer block goes on each side of the mold along the back-up plate. The length of the spacer block is the same length of the back-up plate. The height of the spacer block must allow sufficient space for the traveling of the ejector plate up to the necessary stroke to release the plastic parts from the mold. The width of the spacer blocks must give the required strength to support the compression forces applied to the mold during the injection process. In addition to the spacer blocks, support pillars are used to increase the strength of the mold.

6.8- Support pillars: The support pillars are located between the back-up plate and the rear clamping plate. They are parallel to the spacer blocks reinforcing the core side of the mold structure. The ejector plate has hollow areas to move freely around the support pillars. The amount of support pillars is estimated on a case-by-case basis. They have the same height of the spacer blocks and usually they have a cylindrical shape. The major consideration in establishing the final dimensions of the support pillars is the fitting of these support pillars into the ejector plate area without creating any interference for the lifters, ejector pins and/or any other action located on the ejector plate.

6.9- Rear clamping plate: The clamping plate dimensions were previously explained. It is suggested to adjust the dimensions of the clamping plate in order to fit the injection machine and material supplier standards. This adjustment must meet the technical requirements of the injection machine and the injection mold. The mold clamping plate thickness and steel type may vary case-by-case and they must be defined in order to estimate the final cost of the injection mold.

6.10- Ejector plate: The ejector plate actuates the ejector system (ejector pins, the return pins, and the lifters). The ejector system uses two plates bolted together to hold the components actuated for it. The ejector plate is the rear plate of these two bolted plates. This ejector plate receives the direct actuated force from the injection machine. The ejector plate must fit in between the spacer blocks and the ejector plate estimated thickness is in many cases selected from supplier catalogs.

6.11- Ejector retainer plate: This is the front plate of the ejector system. It is bolted to the ejector plate. It has similar dimensions of the ejector plate and the spring load return system is supported by it. The ejector retainer plate receives the direct force applied from these springs.

6.12- Ejector pins: The ejector pins push the part off of the mold. The bottoms of the ejector pins are enclosed in between the ejector plate and the ejector retainer plate to move them forward and backward with these plates during the ejection of the plastic parts. The head or tips of the ejector pins are in direct contact with the plastic parts. The lengths of these ejector pins are the ejection stroke plus the width of the back-up plate plus the width of the front core plate at the location of the ejector pin. It is very common practice to purchase oversized pins from a supplier catalog and subsequently complete the final fitting later at the tool shop. The locations of the ejector pins, their diameters and other dimensions depend on part shape and tooling configuration.

6.13- Sprue puller: The pullers work exactly like the ejector pins. The purpose of the sprue puller is to pull the runner from the cavity side to the core side during mold opening. The quantities and dimensions depend on runner design and tooling configuration.

6.14- Return pins: The return pins are safety items installed on the ejector plate like the ejector pins. However, they are stronger than the ejector pins. They force the ejector plate to return to the original position when the mold is closing in order to prevent any damage on the cavity side because of action components. These can be purchased from a supplier catalog.

6.15- Leader pins: The leader pins are used to keep the cavity side and the core side of the mold parallel. There are usually four (4) leader pins mounted at the four corners of the mold. When the mold is opened to eject the part, the cavity side and the core side of the mold are pulled apart. The leader pins are the components guiding the cavity side and the core side of the mold when the machine closes in order to keep both sides of the mold centered every single time.

6.16- Leader pin bushing: The leader pin bushings are the guides for the leader pins on the core side of the mold. The tolerances between leader pins and leader pin bushings are very small. They can be purchased together with the leader pin from a supplier catalog.

The following summary table lists the standard components of the injection mold and their critical values in order to estimate the final cost of these items.

Sub-Group	#	Item	Dimensions		Qty.	Price ($/unit)	Total Cost ($)
			Variable	Value			
Steel plates	1	Front clamping plate	X, Y, Z				$0.00
	2	Front cavity plate	X, Y, Z				$0.00
	3	Front core plate	X, Y, Z				$0.00
	4	Support plate (back-up plate)	X, Y, Z				$0.00
	5	Spacer block	X, Y, Z				$0.00
	6	Rear clamping plate	X, Y, Z				$0.00
	7	Ejector plate	X, Y, Z				$0.00
	8	Ejector retainer plate	X, Y, Z				$0.00
Components	9	Support pillars	OD, L				$0.00
	10	Stop pins	OD, L				$0.00
	11	Leader pin	OD, L				$0.00
	12	Leader bushing	ID, L				$0.00
	13	Locating ring	OD				$0.00
	14	Sprue bushing	OD, L				$0.00
	15	Springs for ejection system	ID, L				$0.00
	16	Rod for ejector system	ID, L				$0.00
	17	Parting line locks	X, Y, Z				$0.00
	18	Water connectors	ID				$0.00
	19	Eye bolts	OD, L (#)				$0.00
Pins	20	Sprue puller	OD, L				$0.00
	21	Ejector pins	OD, L				$0.00
	22	Sleeve pin	OD, ID, L				$0.00
	23	Core pin	OD, L				$0.00
	24	Return pin	OD, L				$0.00
Action	25	Total lifter cost (std parts)	units				$0.00
	26	Total Slide cost (std parts)	units				$0.00
		Total standard component cost ($)					$0.00

CHAPTER 7

PARTS MANUFACTURED BY THE MOLD MAKER

The parts manufactured by the mold maker take in consideration all of the components of the injection mold, which are manufactured at the mold shop. The part shape is transferred to the front mold plate through out the machining of the front core plate and front cavity plate of the mold. Any component involved in the shaping of the plastic part must be machined at the mold shop. In addition, the injection mold requires the machining of different areas for functional applications and for the fitting of standard components into the mold. Some of the most common areas to consider at this time are the manufacturing of the cooling system, the holes for the eye bolts, the housing for the leader pins, the housing for the pushing pins, holes for placing the ejector pins, the pocket for sprue bushing, the holes for return pins, the machining of the runners and gates.

Toolmakers may approach the manufacturing of injection molds in many different ways. New technologies, reliable equipment, and qualified personnel are relevant considerations during the layout of a cost effective action plan for the manufacturing of an injection mold.

The sub division of the parts manufactured by the toolmaker into three (3) major groups simplifies the cost estimation activity. Group 1 includes all the machined components in direct contact with the final plastic part. Group 2 considers any additional manufacturing processes that do not have direct contact with the plastic part, and group 3 considers all of the additional activities necessary to complete the injection mold not included in group 1 and/or group 2.

7.1- Group 1: The major components to take in consideration are the front cavity plate, front core plate, and actions such as lifters, and slides.

7.1.1- Front cavity plate: It is very common to use the front cavity plate to shape the "A" surface of the plastic parts. This "A" surface for the final plastic parts requires the specification of finishing, texturing and/or any other necessary process in order to give the finished appearance to the plastic parts. It is very important to specify the required finishing on the surface of the mold to estimate its final cost.

The manufacturing of the front cavity plate and the front core plate is broken down into basic machining and polishing activities. Then, any additional work must be considered and added to the final cost.

The finishing level on the cavity affects the required timing during the machining and the polishing process. The following is a general cost estimation for the manufacturing of the front cavity plate and the front core plate:

As = Surface area of cavity (mm^2)

Ap = Projected area of cavity (mm^2)

R = Average removal rate (mm/h)

Rd = Depth for average removal rate (mm)

Cd = Depth to remove from cavity due to part thickness (mm)

Tc = Timing of removal per cavity (h)

Tc (h) =   (Cd/Rd) [(Ap * R) + (As-Ap)* (0.5 * R)]

Mc ($/h) = Machine and labor cost ($/h).

Cc ($) = Cavity machining cost.

Cc ($) = Mc ($/h) * Tc (h).

This calculation estimates the cost for the manufacturing of the part shape on the cavity of the mold. It is very important to highlight the degree of estimation of this calculation. An error factor may be considered to adjust the unknown element during the machining process of the mold.

Machining time for the cavity of the mold (Rough cut)			
Description		Unit	Value
Mold info.	Surface Area of cavity	As (cm^2)	
	Projected area of cavity	Ap (cm^2)	
	Cavity depth of the part normal to projected area	Cde (cm)	
Values for steel 30-40 HRc (P-20 or similar)	Average removal rate (table speed)	Vf (mm/min)	
	Depth of removal (axial depth 10% of D)	ap (mm)	
	Effective diameter	De (mm)	
	Cutter diameter	D (mm)	
	Rotation speed of the cutter	RPM (rev/min)	
	Chip load per tooth	Fz (mm/unit)	
	Number of tooth or flutes on the cutter	Zn (units)	
	Remove surface per minute	Ra (mm^2/min)	
	Volume removal rate	Qr (mm^3/min)	
	Volume to remove from cavity	Qc (mm^3)	
	Machining time (100 % efficiency)	Tc (h)	
	Machining time (85 % efficiency)	Tcr (h)	

$$De = 2 * (ap\,(D\text{-}ap)\,)^\wedge \tfrac{1}{2}$$

$$Fz = (D*Hex) / De$$

$$Vf = RPM * Fz * Zn$$

$$Ra = Vf * 0.4 * De$$

$$Qr = Ra * ap$$

$$Qc = As * Cde$$

$$\textbf{Tcr (h)} = \textbf{(Qc / (Qr * 60))}$$

Machining time for the cavity of the mold (Finishing)			
	Description	Unit	Value
Mold info.	Surface Area of cavity	As (cm^2)	
	Projected area of cavity	Ap (cm^2)	
	Cavity depth due to Part thickness	Cd (mm)	
Values for steel 30-40 HRc (P-20 or similar)	Average removal rate (table speed)	Vf (mm/min)	
	Depth of removal (axial depth 10% of D)	ap (mm)	
	Effective diameter	De (mm)	
	Cutter diameter	D (mm)	
	Rotation speed of the cutter	RPM (rev/min)	
	Chip load per tooth	Fz (mm/unit)	
	Number of tooth or flutes on the cutter	Zn (units)	
	Remove surface per minute	Ra (mm^2/min)	
	Volume removal rate	Qr (mm^3/min)	
	Volume to remove from cavity	Qc (mm^3)	
	Irregular contour for machining	Qd (mm^3)	
	Machining time (100 % efficiency)	Tc (h)	
	Machining time (85 % efficiency)	**Tcf (h)**	

$$De = 2 * (ap * (D-ap))^{\wedge} \tfrac{1}{2}$$

$$Fz = (D * Hex) / De$$

$$Vf = RPM * Fz * Zn$$

$$Ra = Vf * 0.4 * De$$

$$Qr = Ra * ap$$

$$Qc = As * Cd$$

$$Qd = (As - Ap) * Cd$$

Tcf (h) = (Qc / (Qr*60)) * ((0.5 * Qd) / (Qr * 60))

7.1.2- Front core plate: The front core plate of the mold is used to shape the backside of the plastic part. In addition, this side of the mold is used for the ejection of the plastic part from the mold. The cost estimation for manufacturing the part shape on the front core plate is very similar to the manufacturing of the part shape on the front cavity plate. The major variation from the front cavity plate is the finishing level on the surface of the front core plate. The backside of the plastic part may not need polishing and/or any other level of finishing. The additional cost for reaching a specified finishing level could be eliminated from the manufacturing process at the front core plate of the mold.

7.1.3- Action elements: The action elements are used to shape the plastic part in the mold when an undercut feature is required on the plastic part. This section of the plastic part cannot be formed and ejected through a normal mold opening operation. Action elements shape the part during injection, and then they move away from the undercut area releasing the part from the mold. These action elements are very common on the tooling design. There are two main components of the action elements. The first component (action block) shapes and releases the plastic part from the undercut in the mold. The second component is responsible for moving the component (action block) away from the under cut area during the molding process. A portion of the action block is in direct contact with the part, and it is manufactured by the toolmaker. The rest of the action system (second component) can be purchased from a supplier catalog. Lifters and slides are examples of these action elements and they are the most widely used in the industry.

TOTAL LIFTER COST ($)

	ITEM	Description	Qty.	Price ($)/unit	Total cost
		Qty. of lifters			
MATERIAL	Graphite electrode for EDM	Dimensions (x, y, z)			
	Lifter block (H-13)	Dimensions (x, y, z)			
		TOTAL MATERIAL COST =			

	GROUP	Description	Time	$/hr	Total cost
MANUFATURING	Electrode	Set up cost			
		Machining cost (Milling)			
	EDM (rate 100 mm^3/mim)	Set up cost			
		Electric discharge machining cost (EDM)			
	Machining lifter block	Set up cost			
		Machining cost (Milling)			
	Machining core for lifter housing	Set up cost			
		Machining core for lifter housing (Milling)			
	Machining core for lifter rod housing	Set up cost			
		Lifter rod housing (Drilling gun)			
	Machining ejetor plate	Set up cost			
		Lifter slide base (milling)			
		Total lifter manufacturing time =	0	Total	$0.00

	ITEM	Description	Qty.	Price ($)/unit	Total Cost
STANDARD COMPONENTS	Lifter rod	Supplier catalog			
	Lifter slide plate	Supplier catalog			
	Guide Rails for lifter slide core	Supplier catalog			
	Lifter Slide core (steel) - (Misumi)	Supplier catalog			
	Intermedium bushing for lifter rod	Supplier catalog			
		Total lifter manufacturing time =	0	Total	$0.00

Total lifter cost ($)= $0.00

7.2- Group 2: The injection mold requires the machining of different areas for functional applications and for the fitting of the standard components into the mold. This group takes in consideration manufacturing of all these elements. These elements do not have direct contact with the plastic part, however they have critical functions and they must be included on the mold.

The most common machining activities are the manufacturing of the cooling system, the holes for the eye bolts, the housing for the leader pins, pushing pins, ejector pins, sprue bushing, return pins, runners, and gates.

7.2.1- Cooling system: Fluid is circulated throughout the cooling channels inside the mold in order to remove the heat that has been transferred from the melted plastic inside the cavity to the mold.

During this heat exchange process, the heat from the melted plastic is transferred to the mold where this fluid takes the heat away from the mold and transports it to an external unit where it is dissipated into the atmosphere. This works like a radiator from an internal combustion engine.

Many different fluids can be used for this purpose. However, water is the most widely used cooling fluid in the injection molding industry.

The cooling system controls the mold temperature and it has a direct effect on the quality of the final product and the cycle time during the injection molding process.

The basic process to manufacture these channels is to drill all across from one side to the other of the front cavity plate and the front core plate. These channels are parallel to the surface of the part. The final diameter and the most efficient location for these channels require additional studies. In some cases, additional cooling elements are used to access specific areas and/or increase the cooling time of the part. The cost estimation of straight drilling is acceptable for simple parts. Complex part shapes should add up to 50% of the original estimated cost.

Drilling is a term used to cover all method of making cylindrical holes in a work piece with chip cutting tools.

An extensive bibliography for metal cutting techniques and manufacturing parameters for drilling into steel is commercially available. The following values are suggested for a cost estimation when data is not available for drilling these water channels into the steel. An average speed rate of 0.001 revolutions per minute (rpm) and a cutting speed of 60 feet per minute (fpm) should be used. These values are based on material properties in comparison to P-20, H13.

Also, it is recommended to use water channel diameters equal to or smaller than the waterline connectors specified on the injection molding machine. A water line diameter of 3/8 inches is suggested for cost estimation purposes when data is not available.

Using a cutting speed of 60 feet per minute, and a cutter diameter of 3/8 inches it is possible to determine the revolutions per minute (RPM) for the drilling of these channels (n=611 RPM).

The injection molding machines normally have a manifold with four (4) incoming and four (4) outgoing water connectors per each side of the mold. This information is used to specify eight (8) cooling channels per each side of the mold. The following calculation presents the cost estimation for making the cooling channels (Cch) for a plastic injection mold:

Tch = Machining time for the cooling system (hr).

Dch = Diameter of the drill or the channel (in).

f = Feed rate (in/rev.)

r = RPM of the cutter (rev/min.)

L = Length of the channel (in)

n = Number of channel.

Tch = (n * L) / (f * r*60)

Tch = (8 * L) / (0.001 * 611*60) = 8*L / (0.611*60) = 13.1 * L / 60

Tch (Hr) = 0.22 * L

Mch ($/h) = Machine and labor cost ($/h).

Cch ($) = Cooling channels machining cost.

Cch ($) = Mch ($/h) * Tch (h).

7.2.2- Leader pins and leader bushing:

The halves of the mold themselves are guided internally in order to obtain the necessary accuracy for the sealed off areas at closing.

The injection mold uses the leader pins to accomplish this activity. The leader pins protrude from one mold half of the opened mold and slide into precisely fitting bushings in the other mold half during mold closing. This ensures a constant and accurate alignment of both surfaces of flat molds without shifting during injection and the production of molding. Four leader pin sets (pin and bushing) are normally used for this alignment.

Effective alignment is possible only if close tolerances are kept between leader pins and corresponding holes. Replaceable leader bushings are mounted on the corresponding holes in order to control the tolerances and localize wear between these two components (leader pins and leader bushing). The replaceable leader bushings enable worn-out parts to be exchanged very easily. Leader pins and leader bushings are available commercially.

The corresponding housings to fit these components are drilled to the specified tolerances in the plates of the mold. The cost estimation for drilling the leader pin housings and the leader bushing housing are similar to the previous calculations presented during the cost estimation for the cooling system. However, a machining factor (k) of 1.35 considers the additional time to achieve the necessary accuracy for these areas.

The analysis to determine the cost estimation for drilling the leader pin bushing (Cl) housings is shown below:

Tl = Machining time for the cooling system (hr).

Dl = Diameter of the drill or the leader bushings (in). (from ¾" to 3")

f = Feed rate (in/rev.)

r = RPM of the cutter (rev/min.)

L = Length of the leader bushing (in)

n = Number of leader bushings.

K = Finishing factor

Using the biggest diameter available commercially for leader pins to estimate the cost for drilling these housings is a valid consideration if the leader pins diameter is an unknown value during the cost estimation stage.

It is recommended to specify a standard diameter for the leader bushing. Three (3) inches diameter is suggested for this purpose.

Considering an average cutting speed of 60 fpm, and a cutter diameter of 3 inches it is possible to determine the RPM for the drilling these housings (n=76.4 RPM).

$Tl = (k * n * L) / (f * r*60)$

$Tl = (k * 4 * L) / (0.001 * 76.4 * 60) = (1.35 * 4 * L) / (0.0764*60) = 1.178 * L$

$Tl (Hr) = 1.18 * L$

The length of the leader bushing depends on its diameter. It should be two to three times the leader bushing inside diameter.

$Tl (Hr) = 1.18 * L = 1.18 * (2 * D)$

Ml ($/h) = Machine and labor cost ($/h).

Cl ($) = Leader bushing housing machining cost.

Cl ($) = Ml ($/h) * Tl (h)

    7.2.3- Ejector pin: The ejector pins are the rods that push the molded part out of the mold. The ejector pins are installed on the ejector plate. The ejector pins require an open channel through the core side of the mold in order to allow these pins to pass through and move freely with the ejector plate. These open channels are drilled through the core side of the mold to fit the ejector pin. The ejector pins are commercially available in many different dimensions. The choice of ejector pins is largely governed by article shape, and by the rigidity or flexibility of the raw plastic material used during molding.

The calculation of the required force to strip a molding off a male core is beyond the scope of this study and its consideration will increase the complexity of the analysis for the cost estimation of the plastic injection mold.

There is not a simple straightforward calculation to estimate the amount of ejector pins necessary to eject the part from the male half (core side) effectively. The expertise and judgment of the estimator is required for this purpose. It is recommended to place ejector pins at those areas where the sidewalls are parallel to the direction of the mold opening. Ribs, under cuts, and sidewalls are areas with major concerns during the ejection process.

The cost estimation for drilling the mold has previously been explained in detail. The ejector pins are perpendicular to the surface of the molded part and they move on the direction of the mold opening.

The cost estimation for making the housing for the ejector pins is (Cj) the following:

T_j = Machining time for making the ejector pin housing (hr).

D_j = Diameter of the drill or the ejector pins (in). (From 1/8" to 1")

f_j = Feed rate (in/rev.)

r_j = RPM of the cutter (rev/min.)

L_j = Length of the ejector pin (in)

S = Stroke of the ejector system (in)

n_j = Number of ejector pins.

The required length for drilling the ejector housing is determined by L-S. It is suggested to use the biggest diameter commercially available for leader pins to estimate the cost for drilling these housings. The shape of the part, the rigidity and/or flexibility of the plastic raw material are taken in consideration by the mold designers in order to determine the diameters of the ejector pins. The ejector pins can be purchased by catalog and the housing of the ejector pins must be drilled into the movable side of mold for the mold maker.

The recommended ejector pin diameter for cost estimation purpose is 1/2 inch.

The estimated RPM for drilling the ejector pin housing is calculated using a cutting speed of 60 fpm, and a cutter diameter of 1/2 inch (n=459 RPM).

$Tj = (n * (L-s)) / (f * r*60)$

$Tj = (n *(L-s) / (0.001 * 459 * 60) = (n * (L-S)) / (0.459*60) = 0.036 * n (L-S)$

Tj (Hr) = 0.036 * n (L-S)

Mj ($/h) = Machine and labor cost ($/h).

Cj ($) = Ejector pin housing machining cost.

Cj ($) = Mj ($/h) * Tj (h).

7.2.4- Sleeve pins: The sleeve pins are hollow rods used to create the bosses on the backside of the plastic parts. The sleeve pin is hollow with a fixed core tied to the back plate passing through the hollow sleeve pin up to the part. The core pin shapes the inside diameter of the boss and the sleeve pin shapes the outside diameter of the boss. During the ejection process the sleeve pin moves forward with the ejector plate releasing the part from the fixed core pin. The sleeve pin and the core pin are commercially available, and can be purchased as a set. The quantity of sleeves and core pins is equal to the number of bosses on the molded part.

The diameter for the housing of the sleeve pins on the core side of the mold is equal to the outside diameter of the sleeve pin, and the diameter of the sleeve pin is equal to the outside diameter of the plastic boss on the part.

The following is the cost estimation analysis of making the housing of the sleeve pin on the core side of the mold:

Ts = Machining time for making the sleeve pin housing (hr).

Ds = Diameter of the sleeve pin (in). (From 3/16" to 1")

fs = Feed rate (in/rev.)

rs = RPM of the cutter (rev/min.)

Ls = Length of the sleeve pin (in)

S = Stroke of the ejector system (in)

ns = Number of sleeve pins.

The required length for drilling the sleeve pin housing is determined by L-S. The sleeve pins can be purchased by catalog, and the housing for these pins must be drilled into the movable side of mold.

If the outside diameter of the boss is not available during the cost estimation activity, an outside diameter of 3/8 inches for the sleeve pin is a reference dimension that the cost estimator could use.

Using a cutting speed of 60 feet per minute, and a cutter diameter of 3/8 inch. It is possible to determine the RPM for the drilling of these housings (n= 611 RPM).

$$Ts = (n * (L\text{-}s)) / (f * r*60)$$

Ts = (n *(L-s) / (0.001 * 611 * 60)=(n * (L-S)) /

(0.611*60) = 0.028 * n (L-S)

Ts (Hr) = 0.028 * n (L-S)

Ms ($/h) = Machine and labor cost ($/h).

Cs ($) = Sleeve pin housing machining cost.

Cs ($) = Ms ($/h) * Ts (h).

7.2.5- Return pins: The return pins ensure
the return of the advanced ejector plate to its rest
position during the closing of the mold. Otherwise
the ejection assembly or the opposite mold half
may be damaged. The return pins can be used as
the primary return system for the injection mold or
to provide a safeguard for return of the ejector
plate. Therefore, a combination of a return spring
and a return pin is frequently used. It is a common
practice to place four (4) return pins on small and
medium size molds on each corner of the ejector

plate, and the number of them will increase as needed for bigger size molds.

The return pins must be strong enough to push the ejector plate back to the returned position. The diameter of the return pins depends on the required force used to push the ejector plate back to returned position. The return pins can be purchased from a supplier catalog. Common diameters for return pins range from ½" to ¾". For cost estimation purposes four (4) return pins with a diameter of ¾" are recommended.

The following analysis presents the cost estimation for drilling the housings of the return pins (Cr):

Tr = Machining time for making the return pin housing (hr).

Dr = Diameter of the return pin (in). (From 3/16"
to 1")

fr = Feed rate (in/rev.)

rr = RPM of the cutter (rev/min.)

L = Length of the return pin (in)

S = Stroke of the ejector system (in)

nr = Number of return pins.

The required length for drilling the return
pin housing is determined by L-S. The suggested
diameter for the return pins during the cost
estimating process is ¾" and the number of pins
are four (4). The return pins and the housing for
these pins must be drilled into the movable side of
the mold.

The estimated (RPM) for drilling the return
pin housing is calculated using a cutting speed of

60 fpm, and a cutter diameter of 3/4 inch (n=306 RPM).

Tr = (n * (L-s)) / (f * r*60)

Tr = (n *(L-s) / (0.001 * 306 * 60)=(n * (L-S)) / (0.306*60) = 0.055 * 4 (L-S)

Tr (Hr) = 0.22 (L-S)

Mr ($/h) = Machine and labor cost ($/h).

Cr ($) = Ejector pin housing machining cost.

Cr ($) = Mr ($/h) * Tr (h).

7.2.6- Sprue puller: The sprue puller is a rod used to pull or draw a molded sprue out of the sprue bushing; it is generally a straight round pin with the end machined in the form of an undercut. It is fastened to the ejector mechanism and runs through the movable part of the mold in direct line with the mold entrance.

The sprue puller is perpendicular to the surface of the part and moves in the direction of the mold opening. The following is the cost estimation for drilling the housing of the sprue puller (Csp).

Tsp = Machining time for making the ejector pin housing (hr).

Dsp = Diameter of the drill or the sprue (in). (3/8")

fsp = Feed rate (in/rev.)

rsp = RPM of the cutter (rev/min.)

L = Length of the sprue bushing (in)

S = Stroke of the ejector system (in)

nsp = Number of sprue puller.

The required length for drilling the sprue puller housing is determined by L-S. The diameter of the sprue puller is proportional to the diameter of the sprue bushing at the cavity of the mold. And the required undercut must be strong enough to remove the molded sprue from the cavity side of the mold. The size of the sprue, and the rigidity and/or flexibility of the plastic material are used to determine the final diameter of the sprue puller. The sprue puller can be purchased from a sub supplier catalog, and the housing of the sprue puller is drilled into the movable side of mold.

One (1) sprue puller of 3/8 inches diameter is suggested for this calculation.

The estimated RPM for drilling the ejector pin housing is calculated using a cutting speed of 60 fpm, and a cutter diameter of 3/8 inch (n=611 RPM).

$Tsp = (n * (L-s)) / (f\ sp* rsp * 60)$

Tsp = (n *(L-s) / (0.001 * 611 * 60)=(1 * (L-S)) /

(0.611*60) = 0.0273 * (L-S)

Tsp (Hr) = 0.036 * (L-S)

Msp ($/h) = Machine and labor cost ($/h).

Csp ($) = Ejector pin housing machining cost.

Csp ($) = Msp ($/h) * Tsp (h).

The following table summarizes the formulas for all the components of the group 2.

Group 2							
#	Items	Formula	Variables	Value	Tc (hr)	Mch ($/hr)	Cch ($)
1	Cooling system	Tc=n*0.22*L	L (in)				$0.00
2	Leader bushing housing	Ti=2.36*D	D (in)				$0.00
3	Ejector pin housing	Tj=0.036*n*(L-S)	n(unit)				$0.00
			L (in)				
			S(in)				
4	Sleeve pin housing	Tj=0.028*n*(L-S)	n(unit)				$0.00
			L (in)				
			S(in)				
5	Return pin housing	Tj=0.055*n*(L-S)	n(unit)				$0.00
			L (in)				
			S(in)				
6	Sprue puller housing	Tsp=0.036*n^(L-S)	n(unit)				$0.00
			L (in)				
			S(in)				
Total cost - Group 2 ($)							$0.00

7.3- Group 3: This group considers all the necessary activities for the completion of the injection mold not included in group 1 or group 2. The activities from this group have direct impact in the final cost and quality of the injection mold. The most relevant activities in this group are mold design, programming, machine set up for manufacturing of the mold, finishing, spotting, fitting, assembly, transportation, trials, and processing fee(s).

7.3.1- Mold design: The mold makers could approach the mold design in many different ways. However, the tooling design should start after the tool shop has received the 3D model of the plastic part (three-dimensional data), 2d drawing of the plastic part, and the mold specification sheet.

The 2d drawing of the part (two-dimensional part print) shows the part specifications (material, tolerances, etc.), and the mold specification sheet defines the characteristics of the mold.

The mold design requires a review of the preliminary mold layout, and a final mold approval previous to proceeding with the manufacturing of the injection mold.

The preliminary mold layout shows the overall sizes of the mold in order to approve the mold maker for ordering the steel. In addition, this review is to request any necessary adjustment on the mold design previous to starting with the manufacturing of the mold.

Final Approval: The final mold design is reviewed and approved after the adjustments have been made. This final mold design approval gives the mold maker the go ahead to start cutting steel.

The required time for mold design depends on the complexity of the mold and the resources allocated to this activity. The preliminary mold layout may take from five days up to 14 days. The final approval requires a lot more considerations and in some cases partial approval is awarded to start the manufacturing of the mold. In some cases partial approvals are required in some locations of the mold previous to cutting steel on those areas. This activity may take from four days up to five weeks. If there is not enough information about the expected time to complete the design of the tool the following information is the suggested reference to follow:

A- For small size molds (injection machines 110 tons or less):

Preliminary design (Pd): 3 days

Final design (Fd): 5 days

B- For medium size molds (injection machines between 110 tons and 650 tons):

Preliminary design (Pd): 5 days

Final design (Fd): 10 days

C- For big size molds (injection machines 650 tons or more):

Preliminary design (Pd): 6 days

Final design (Fd): 15 days

Note: This estimation considers 8 hours per day as working time.

Td (t) = Designing time

Td (Hr) = (Pd + Fd) *8

Md ($/h) = Machine and labor cost ($/h).

Cd ($) = Designing cost.

Cd ($) = Md ($/h) * Td (h).

Some mold makers are still designing the mold in 2d (two dimensional drawings). Then, they use these 2d drawings for the manufacturing of the mold.

Other tool shops use the latest technology in design and manufacturing. These mold makers use computer software to make the design of the mold in a 3D solid model, and create the cutting path to generate the CNC program, then this information is transferred to the CNC machine for the manufacturing of the mold.

This area of mold design and manufacturing is the most difficult to estimate. The latest technology that gave the mold shop excellent results in a previous job may not be cost effective for the next one. The use of excessive edge technology in some cases increases the overhead of the company making its products more expensive to produce. On the other hand, avoiding the use of new technology will increase the manufacturing cost and risk the quality of the product. Determining the balance between the traditional manufacturing and the technological advances to be implemented in a mold shop in order to maintain its competitiveness is one of the biggest challenges facing the mold makers at this time.

7.3.2- Programming: The major impact of the cnc programming activity is located on the cavity plate of the mold, the core plate of the mold and the action elements such as lifters and slides responsible for the final shape of the molded plastic part(s). As a general rule any element responsible for shaping plastic part(s) requires programming for the manufacturing process.

The complexity of the programming activity depends on part shape, required finishing, specified tolerances, and machine capabilities, among others.

During the cost estimation stage the cnc programming time could be considered proportional to the machining time for a specific part. It is suggested to estimate factor (Kp) ranging between 20% and 35% of the machining time for programming activities. This programming time is additional to the machining time and it should be considered separate.

The following shows the cnc programming cost break down analysis:

Tc1 = Time for machining the cavity

Tc2 = Time for machining the core

Ts = Time for machining the slides

Tl = Time for machining lifters

Kp = Factor from machining time to programming time

Tp = Timing for programming

Tp = Kp (Tc1+Tc2+Ts+Tl)

Mp ($/h) = Machine and labor cost ($/h).

Cp ($) = programming cost.

Cp ($) = Mp ($/h) * Tp (h).

7.3.3- Benching: Benching is usually a manual activity that requires the use of power and manual hand tools to prepare the surface to be polished and complete the tuning or adjustment for the final fitting of the matting parts of the mold. The estimated time for this activity is determined by the size of the mold. The size of the mold is proportional to the size of the injection machine where it is going to be used. The injection molds are classified into three groups: small, medium, and large indicating the size of the injection molding machines.

Small size molds= injection machines 110 tons or less

Medium size mold= injection machines between 110 and 650 tons

Large size mold= injection machines 650 tons or more

Benching (time):

T_{bs} = Small size molds: 2 day (8 hrs)

T_{bm} = Medium size molds: 3 day (24 hrs)

T_{bL} = Large size molds: 5 day (40 hrs)

Tbac = Actions (time per each action): 3 hrs (any size)

Na= Number of actions.

Ta = (Tbs, Tbm, or TbL) + Na * Tbac

Ta= Benching time.

Ma ($/h) = Machine and labor cost ($/h).

Ca ($) = Assembly cost.

Ca ($) = Mc ($/h) * Tc (h).

   7.3.4- Spotting: The spotting of the mold ensures an effective shut off around the parting lines to prevent any flash during the molding process. The most popular technique to spot a plastic injection mold is mounting the injection mold on a press, opening the mold and covering one of the sides of the mold with a colorant film. Then, the mold is closed to simulate shut off condition around the parting line. The mold is opened to verify the contact of shut off areas and ensure the correct seal off around the parting line of the mold. This activity is repeated and the mold is reworked until all the shut off areas are sealed completely.

The following presents the cost estimation for spotting the mold:

Small size molds: injection machines 110 tons or less

Medium size mold: injection machines between 110 and 650 tons

Large size mold: injection machines 650 tons or more

Spotting (time):

Tss = Small size molds: 4 hrs

Tsm = Medium size molds: 6 hrs

TsL = Large size molds: 8 hrs

Tsac = Actions (time per each action): 2 hrs (any size)

Na= Number of actions

Tb = (Tss, Tsm, or TsL) + Na * Tsac.

Tb= Spotting time.

Mb ($/h) = Machine and labor cost ($/h).

Cb($) = Spotting cost.

Cb ($) = Mc ($/h) * Tb (h).

7.3.5- Assembly: The assembly of the mold is the activity where all the components of the mold are put together. Every single component is assembled or sub-assembled in order to have the injection mold as a complete product. The required time for this activity depends on the availability of all the components, and the procedures established for the mold maker to optimize this activity.

Tas = Small size molds: 4 hrs

Tam = Medium size molds: 6 hrs

TaL = Large size molds: 10 hrs

Tc = (Tas, Tam, or TaL)

Tc= Assembly time.

Mc ($/h) = Machine and labor cost ($/h).

Cc($) = Assembly cost.

Cc($) = Mc ($/h) * Tc (h).

   7.3.6- Polishing: Plastic molds usually require a high level of polishing on the "A" surface. A large part of the mold cost is involved with this activity. The polishing of the mold represents from five percent (5 %) to 30% of the mold cost. One must recognize that an experienced polisher could polish from 2 to 5 sq in./h (5.1 to 12.7 sq cm./h) by hand. Certain shops can at least double this rate if they have the right equipment.

The amount of polishing on the mold is directly proportional to the quality of the cutting path during the machining stages and the expected finishing level for the injection mold.

The part drawing specifies the finishing level expected for the plastic part and its locations. This information is used to define the required polishing for the injection mold in order to transfer the expected finishing level to the part.

Selecting a higher level of polishing than it is actually needed significantly increases the final cost of the mold.

The No-3 finish is installed on the cavity by finishing with a 320 grit abrasive cloth. This fine finish level can be used for post operations such as texturing and plating. The core side of the mold in many cases is not on the "A" surface of the plastic part. Since this is not visible, a lower level of polishing or no polishing at all is acceptable. This reduction or elimination of the polishing on the area of the part not visible reduces the cost of the tool significantly.

7.3.7- Transportation: The freight cost includes the transportation of the mold from the mold shop to the injection molding facility or any other location specified on the contract. The transportation cost increases proportionally to the weight of the mold, and the distance and method used for transporting the mold. This transportation cost does not consider any other freight such as the shipping of assembly parts of the mold or transportation of the mold for tooling trials during the building of the mold. Any additional cost like air shipment must be specified in advance, and further considerations are required.

The following is the required information to determine the cost for the transportation of the mold by a truck:

Distance of shipping (miles): This is the distance between tool shop and molder facility.

Weight of the mold (Tons): For the selection of a truck capable of transport the mold.

Mold size (X,Y,Z): External dimensions of the mold

Truck type: single bed or double bed truck.

The freight cost is verified throughout different transportation companies. These prices are normally flat rates, and they are very easy to obtain when the distance of shipment, the weight of the mold, and the external size of the mold are known values.

7.3.8- Mold trials: The mold trials are used to verify the capability of the mold for making plastic parts within the specifications. It is recommended that the cost estimator considers at least two (2) or three (3), injection mold trials in the final price of the mold. The first mold trial is usually to verify the mold functioning, and review the plastic parts. Second and third molding trials are very important to complete any necessary adjustment on the mold in order to achieve good quality product(s) in an efficient manner.

The cost for a molding trial is based on the size of the injection machine required for molding the parts, time spent during the trial, and the plastic material used during the mold trial. The injection machine is specified on the mold specification sheet, and the plastic material is specified on the part specification sheet.

The time spent during a mold trial takes in consideration the time for putting the mold on the injection machine, the time for adjusting the machine setup for the new mold, and any additional activity involved with the mold trial. The use of flat rates considering the size of the machine is a very common practice. Also, an injection mold trial should consider the making of 50 or 100 plastic pieces when it is not specified.

Small machine (less than 110 tons): $ 600 per trial

Medium machine (from 110 tons to 650 tons) : $ 750 per trial

Big machine (650 tons or bigger): $ 1,000 per trial.

7.3.9- Processing fee(s): The processing fee(s) is the cost incurred for keeping the project up to date. Planning schedules, progress reports, presentations, and mold review meetings are required activities to coordinate and track the progress of the injection mold.

The processing fee(s) represents the required work for planning, maintaining, and completing the project in order to meet the customer expectations in a timely manner.

The amount of activities required among customers and each particular mold maker is very difficult to anticipate. It is suggested that the cost estimator should consider between five percent (5%) and ten (10%) of the total mold cost for processing fees. Some toolmakers hide this cost in the manufacturing of the mold. Others consider so many items that may influence the outcome of the quote significantly.

The following is the equation recommended for the cost estimation of the processing fees:

Processing fee ($) = Total mold cost ($) * 0.10

7.3.10- Additional Items: The additional items for the completion of a plastic injection mold vary on a case-by-case basis. These additional items have particular applications for the plastic injection mold. They should be calculated and added to the total cost of the mold. Some of the most common additional items to take in consideration are the textures of the mold, hot runner systems, coating of the mold, and gas assisted equipments.

The following table summarizes the items presented in group 3 of the capital budgeting for a plastic injection mold for the part manufactured by

the mold maker in this analysis.

						Mch	
#	Item	Description	Variables	Value	Tc (hr)	($/hr)	Coh ($)
			Group 3				
1	Mold Design	Small mold=64	hr				$0.00
		Medium mold=120	hr				
		Big mold=168	hr				
2	Programing	Tp=(Tc1+tc2+Ts+Ti)*Kp	Tc1				$0.00
			Tc2				
			Ts				
			Ti				
			Kp				
3	Benching	Small mold=16	hr				$0.00
		Medium mold=24					
		Big mold=40					
		Time per action	hr/units				
		Qty. of slides	units				
		Qty. of lifters	units				
4	Spotting	Small mold=4	hr				$0.00
		Medium mold=6					
		Big mold=8					
		Time per action	hr/units				
		Qty. of slides	units				
		Qty. of lifters	units				
5	Polishing	Polishing rate = 120	mm^2/h				$0.00
		Surface Area of cavity (As)	mm^2				
6	Assembly	Small mold=4	hr				$0.00
		Medium mold=6					
		Big mold=10					
7	Mold Trial	Small mold=600	$/trial				$0.00
		Medium mold=650					
		Big mold=1,000					
		Number of mold trials	trial				
8	Transportation	Distance	miles				$0.00
		Mold weight	lb				
		Truck rate	$				
9	Processing fees	Total mold cost	$				$0.00
		(%) of Processing fees (10%)	%				
10	Additional items	Hot runner systems, pressure sensors, water manifolds, etc	Units				$0.00
		Total Cost - Group 3 ($)					$0.00

CHAPTER 8

TOTAL COST
OF THE INJECTION
MOLD

The cost estimation for a plastic injection mold is a multidisciplinary activity that requires the direct involvement of different departments in an organization. The sales department must define the customer expectations for the final plastic product and the projection of the production forecast during the life of the program. This information is critical for the engineering group in order to develop a tooling specification sheet for the building of an injection mold capable of meeting the customer's expectations throughout the life of the program.

The cost estimation department breaks down the elements from the tooling specifications sheet into single groups in order to price each activity and each component required for the building of the injection mold. The final estimated cost for an injection mold is considered in the initial investment of a project. The financial department uses this information to layout the different financial decision making models in order to determine the impact of the investment on the shareholder's wealth.

This book gives special attention in closing the technical gap between the purchasing and the engineering departments in an organization. This allows for a more smooth transition in the building of a plastic injection mold.

This analysis begins with the definition of the customer expectations for final plastic molded parts and ends with a final cost for plastic injection molds, which can be used to determine the acceptance or rejection of any particular project.

This cost estimation methodology is designed to guide the personnel involved with this activity through the necessary sequence of events to determine the final price of an injection mold capable of meeting customer expectations. The major implication for this research is the consideration of the customer's expectations into the engineering specification for the building of a plastic injection mold. This approach requires the translation of customer expectations for a final plastic product into injection mold specifications that the mold maker is capable of understanding for the manufacturing of the mold in order to meet customer requirements within the budget and in a timely manner.

The purchasing department in an organization is responsible for the acquisition of assets for the company. The injection molds are an important part of the initial investment for any project that requires them. They will impact the quality and cost of the final plastic product during the life of the project.

148

The purchasing personnel are cost oriented, and in many cases they do not understand or, do not have enough time to evaluate the different engineering requirements for a plastic injection mold that fit the particular needs of every project. This situation leaves many of the relevant information for the building of the injection mold in the hands of the mold maker, resulting in the building of the injection mold designed to meet the mold maker's standards for the lowest possible cost. This activity does not take in consideration the customer's expectations and/or the interest of the company placing the order for the injection mold.

This study redefines the complete approach on the decision making for the purchasing of an injection mold from a pure cost saving analysis on the initial investment to a multidisciplinary activity that estimates the actual value of an injection mold capable of meeting the costumer's expectations. This results in a cost-effective investment that ensures the customer's satisfaction and maximizes the shareholder's wealth.

The final cost of the mold represents the sum of the cost of every element of the mold, plus the cost for any additional item or activity involved during the mold building process.

Total cost (\$)= Σ cost1...costn

The total cost of the plastic injection mold determines the net investment for the capital budget analysis of a project. This information could be used to determine the net cash flow during the life of the project and the total net investment. In addition, these are the bases for the development of the capital budget for a plastic injection mold.

FINAL COST OF THE PLASTIC INJECTION MOLD			
VAR	DESCRIPTION	COST ($)	COST ($)
A	Total cost for standard components ($)	A	$0.00
B	Total cost for part manufactured per the mold maker ($)	B	$0.00
C	Processing fees (10% of total cost) ($)	$C = (A + B) \times (0.10)$	$0.00
D	Tolerance of deviation in the cost estimation (10% of total cost) ($)	$D = (A + B + C) \times (0.10)$	$0.00
E	Overhead and unexpected activities (10%)	$E = (A + B + C + D) \times (0.10)$	$0.00
F	Profit (15% of total cost)	$F = (A + B + C + D + E) \times (0.15)$	$0.00
FINAL COST OF THE PLASTIC INJECTION MOLD		$(A+B+C+D+E+F)$	$0.00

CHAPTER 9

CONCLUSIONS AND RECOMMENDATIONS

Conclusions: The practical and theoretical techniques for designing and manufacturing plastic injection molds are achieved with the interrelationship of many different professions. Establishing the engineering requirements for designing and manufacturing plastic injection molds is a complete area of study in and of itself. The literature on this area requires continuous

152

updates. This situation challenges mold builders to determine the right balance between existing and new technology to make a quality mold and stay competitive in the industry.

Purchasing a plastic injection mold is a very complex activity, which requires the consideration of every element that may affect the final price of the mold and/or the quality of the part. The mold specification sheet summarizes all the required elements to be considered for the building of a mold. This sheet reflects the customer's expectations in words that the toolmaker can understand to build a specific mold. In addition, this mold specification sheet is for making a quote. A quote indicates the required cost and timing to build the mold under pre-established customer expectations.

A change on any element of the mold specification sheet may affect the quality or the cost of the mold. One of the biggest challenges for any mold buyer is to specify the required elements needed to build a plastic injection mold thus giving one the ability to make parts that meet the quality requirements during the complete life of a project for the best possible price.

Recommendations: The personnel following this methodology must have the necessary background in this field in order to assertively adjust these basic principles to the particular requirements of a specific plastic injection mold.

This study takes in consideration the up to date existing technology in each one of the different fields required to complete this analysis.

Further research should be considered for this study in order to evaluate and adjust this analysis for any radical changes presented in technology and/or any other area of expertise used to arrive to the final results. This is necessary to maintain the accuracy of these estimated values. The formulas presented for the manufacturing of the plastic injection mold and the selection of its components are real equations used in the industry for these purpose. The use of average values in order to simplify these calculations to predict the final cost of the plastic injection mold limits these results to cost estimation activities only.

CHAPTER 10

SAMPLE

EXERSICE

Estimate the final cost of a plastic injection mold using the following information.

Number of part per unit = 1

Number of parts per year = (Y1=250,000; Y2=280,000;   Y3=220,000;   Y4=220,000; Y5=185,000).

Part dimensions (X, Y, Z) = 72.39 cm, 93.98 cm, 10.16 cm

Surface area of the part "A surface" (cm^2) = 3,700 cm^2

Projected area of the part (cm^2) = 2,808 cm^2

Thickness of the part (mm) = 2.00 mm

Number of Lifters (action part) = 10 units

- Surface area of the undercut (lifter) = 120 mm

- Projected area of the undercut (lifter area) = 100 mm

- Undercut wall thickness (lifter are)= 2.00 mm

Number of slides = (action part) = 1 unit

- Surface area of the undercut (slide area) = 120 mm

- Projected area of the undercut (slide area) = 100 mm

- Undercut wall thickness (slide area) = 2.00 mm

Number of bosses on the part = 12 units,

Transportation cost = $1,200,

Hot runner systems = $28,000

#	Item	Dimensions		Qty.	Price($)/unit	Total Cost
		Variables	Actual Dim.			
			Standard Components			
1	Front Clamping Plate	X, Y, Z	42",46",1.5"	1	1,511.00	1,511.00
2	Front Cavity Plate	X, Y, Z	38", 46", 8.5"	1	6,566.00	6,566.00
3	Front Core Plate	X, Y, Z	38", 46", 8.5"	1	6,566.00	6,566.00
4	Support Plate (Back up plate)	X, Y, Z	38", 46", 1.5"	1	1,500.00	1,500.00
5	Spacer block	X, Y, Z	3", 46", 8"	2	358.00	716.00
6	Rear Clamping Plate	X, Y, Z	42",46",1.5"	1	1,511.00	1,511.00
7	Ejector Plate	X, Y, Z	32", 46", 2"	1	4,130.00	4,130.00
8	Ejector retainer plate	X, Y, Z	32", 46", 1"	1	1,151.00	1,151.00
9	Support pillars	OD, L	3", 8"	12	43.00	516.00
10	Stop Pins	OD, L	5/16", 5/8"	10	9.80	98.00
11	Leader pin	OD, L	2", 10"	4	40.50	162.00
12	Leader bushing	ID, L	2", 3"	4	20.25	81.00
13	Locating Ring	OD	3.99"	1	18.00	18.00
14	Sprue bushing	OD, L	2", 9.5	1	122.50	122.50
15	Springs for ejection system	ID, L	2", 8"	6	23.25	139.50
16	Rod for ejector system	ID, L	2", 9.75"	6	36.50	219.00
17	Parting line Locks(PLF-005)	X, Y, Z	5",1.87",4.51"	4	187.50	750.00
18	Water Connectors	ID	3/8NPT (JP-353)	16	1.00	16.00
19	Eye bolts	OD, L (#)	36mm, 60mm (M36)	8	16.00	128.00
20	Sprue puller	OD, L	0.25", 9.5"	3	5.20	15.60
21	Ejector pins	OD, L	0.5", 9.5"	20	7.85	157.00
22	Sleeve pin	OD, ID, L	6mm,4mm,10"	12	36.00	432.00
23	Core pin	OD, L	4mm,10"	12	3.30	39.60
24	Return pin	OD, L	0.75", 10"	4	9.90	39.60
25	Total lifter cost (Material, manufacturing, STD parts)	units	Lifter	10	3,271.70	32,717.00
26	Total Slide cost (Material, manufacturing,STD parts)	units	Slide	1	5,614.08	5,614.08
27	Additional items		Hotrunner manifold	1	28,000.00	28,000.00
28	Mold trial			3	1,000.00	3,000.00
29	Transportation			1	1,200.00	1,200.00
	Total standard component cost					$97,115.88

Row groups (left margin labels): Steel Plates (items 1–8), Components (items 9–19), Pins (items 20–24), Actions (items 25–26), Others (items 27–29).

Manufactured parts by the mold maker				
	#	Item	Time (H)	Machine and labor cost ($/H)	Total Cost
Primary Group	1	Cavity (Machining)	43	75	$3,227.88
	2	Core (Machining)	43	75	$3,226.39
	3	Actions (Lifters, Slides)	These values are added to the list of std componets.		
Secondary Group	4	Cooling System	2	75	$148.50
	5	Leader Bushing	7	75	$531.00
	6	Ejector pin housing	15	75	$1,134.00
	7	Sleeve pin housing	12	75	$882.00
	8	Return pin housing	7	75	$495.00
	9	Sprue puller housing	1	75	$81.00
Other Group (No manufacturing items)	10	Mold Design	168	75	$12,600.00
	11	Programing	140	75	$10,486.28
	12	Benching	70	75	$5,250.00
	13	Spotting	28	75	$2,100.00
	14	Polishing	31	75	$2,312.50
	15	Assembly	10	75	$750.00
	16	Processing fees			$0.00
Total Manufactured part cost					$43,224.55

Machining time for the cavity of the mold (Rough cut)

Mold info.	Surface Area of cavity	As (cm^2)	3700
	Projected area of cavity	Ap (cm^2)	2808
	Cavity Depth of the part normal to projectea area	Cde (cm)	101.6
Values for steel 30-40 HRc (P-20 or similar)	Average removal rate (table speed)	Vf (mm/min)	1800
	Depth of removal (axial depth 10% of D)	ap (mm)	1
	Effective diameter	De (mm)	6.00
	Cutter diameter	D (mm)	10
	Rotation speed of the cutter	RPM (rev/min)	6,000
	Chip load per tooth	Fz (mm/unit)	0.150
	Number of tooth or flutes on the cutter	Zn (units)	2
	Remove surface per minute	Ra (mm^2/min)	4320
	Volumen removal rate	Qr (mm^3/min)	4320
	Volumen to remove from cavity	Qc (mm^3)	285292.8
	Machining time (100 % efficiency)	Tc (h)	1.10
	Machining time (85 % efficiency)	**Tcr (h)**	**1.27**

$De = 2 \cdot (ap(D-ap))^{1/2}$

$Fz = (D \cdot Hex)/De$

$Vf = RPM \cdot Fz \cdot Zn$

$Ra = Vf \cdot 0.4 \cdot De$

$Qr = Ra \cdot ap$

$Qc = As \cdot Cde$

$Tcr (h) = (Qc/(Qr \cdot 60))$

Machining time for the cavity of the mold (finishing)

	Surface Area of cavity	As (cm^2)	3,700
Mold info.	Projected area of cavity	Ap (cm^2)	2,808
	Cavity Depth due to Part thickness	Cd (mm)	2
Values for steel 30-40 HRc (P-20 or similar)	Average removal rate (table speed)	Vf (mm/min)	600
	Depth of removal (axial depth 10% of D)	ap (mm)	0.05
	Effective diameter	De (mm)	0.30
	Cutter diameter	D (mm)	0.5
	Rotation speed of the cutter	RPM (rev/min)	20,000
	Chip load per tooth	Fz (mm/unit)	0.015
	Number of tooth or flutes on the cutter	Zn (units)	2
	Remove surface per minute	Ra (mm^2/min)	72
	Volumen removal rate	Qr (mm^3/min)	3.6
	Volumen to remove from cavity	Qc (mm^3)	7400
	Irregular contour for machining	Qd (mm^3)	892
	Machining time (100 % efficiency)	Tc (h)	36.32
	Machining time (85 % efficiency)	**Tcf (h)**	**41.77**

De=2* (ap(D-ap))^ 1/2
Fz= (D*Hex)/De
Vf= RPM*Fz*Zn
Ra=Vf*0.4*De

Qr=Ra*ap
Qc=As*Cd
Qd=(As-Ap)*Cd
Tcf (h)=(Qc/(Qr*60))*((0.5*Qd)/(Qr*60))

Total Cavity machining time (Tcf+Tcr) =   43   hr

Machining time for the Core of the mold (Rough cut)

Mold info.	Surface Area of cavity	As (cm^2)	3700
	Projected area of cavity	Ap (cm^2)	2808
	Core Depth of the part normal to projectea area	Cde (mm)	100
Values for steel 30-40 HRc (P-20 or similar)	Average removal rate (table speed)	Vf (mm/min)	1800
	Depth of removal (axial depth 10% of D)	ap (mm)	1
	Effective diameter	De (mm)	6.00
	Cutter diameter	D (mm)	10
	Rotation speed of the cutter	RPM (rev/min)	6,000
	Chip load per tooth	Fz (mm/unit)	0.150
	Number of tooth or flutes on the cutter	Zn (units)	2
	Remove surface per minute	Ra (mm^2/min)	4320
	Volumen removal rate	Qr (mm^3/min)	4320
	Volumen to remove from core	Qc (mm^3)	280,800
	Machining time (100 % efficiency)	Tc (h)	1.08
	Machining time (85 % efficiency)	**Tcr (h)**	**1.25**

$De = 2*(ap(D-ap))^{1/2}$

$Fz = (D*Hex)/De$

$Vf = RPM*Fz*Zn$

$Ra = Vf*0.4*De$

$Qr = Ra*ap$

$Qc = (As*Cde)+(Cde*Ma-Ap)$

$Tcr (h) = (Qc/(Qr*60))$

Machining time for the Core of the mold (finishing)

Mold info.	Surface Area of cavity	As (cm^2)	3700
	Projected area of cavity	Ap (cm^2)	2808
	Core Depth due to Part thickness	Cd (mm)	2
	Average removal rate (table speed)	Vf (mm/min)	600
Values for steel 30-40 HRc (P-20 or similar)	Depth of removal (axial depth 10% of D)	ap (mm)	0.05
	Effective diameter	De (mm)	0.30
	Cutter diameter	D (mm)	0.5
	Rotation speed of the cutter	RPM (rev/min)	20,000
	Chip load per tooth	Fz (mm/unit)	0.015
	Number of tooth or flutes on the cutter	Zn (units)	2
	Remove surface per minute	Ra (mm^2/min)	72
	Volumen removal rate	Qr (mm^3/min)	3.6
	Volumen to remove from core	Qc (mm^3)	7400
	Irregular countour for machining	Qd (mm^3)	892
	Machining time (100 % efficiency)	Tc (h)	36.32
	Machining time (85 % efficiency)	**Tcf (h)**	**41.77**

$De = 2*(ap(D-ap))^{1/2}$

$Fz = (D*Hex)/De$

$Vf = RPM*Fz*Zn$

$Ra = Vf*0.4*De$

$Qr = Ra*ap$

$Qc = As*Cd$

$Qd = (As-Ap)*Cd$

$Tcf (h) = (Qc/(Qr*60))*((0.5*Qd)/(Qr*60))$

Total Cavity machining time (Tcf+Tcr) =   43   hr

162

TOTAL LIFTER COST ($)					
	Qty. of lifters		10		
	ITEM	**Description**	Qty.	Price($)/unit	Total cost
MATERIAL	Graphite electrode for EDM	Dimensions (100mm, 200 mm, 30 mm)-Misumi	1	93.3	$93.30
	Lifter block (H-13)	Dimensions (110, 60 mm , 70 mm)	1	83	$83.00
		TOTAL MATERIAL COST			$176.30

	GROUP	**Description**	Time	$/hr	Total cost
MANUFATURING	Electrode	Set up cost	5	75	$375.00
		Machining cost (Milling)	1	75	$75.00
	EDM (rate 100 mm^3/mim)	Set up cost	8	75	$600.00
		Electric discharge machining cost (EDM)	1	75	$75.00
	Machining lifter block	Set up cost	7	75	$525.00
		Machining cost (Milling)	5	75	$375.00
	Machining core for lifter housing	Set up cost	2	75	$150.00
		Machining core for lifter housing (Milling)	1	75	$75.00
	Machining core for lifter rod housing	Set up cost	2	75	$150.00
		Lifter rod housing (Drilling gun)	1	75	$75.00
	Machining ejetor plate	Set up cost	2	75	$150.00
		Lifter slide base (milling)	3	75	$225.00
		Total lifter manufacturing time =	38	Total	$2,850.00

	ITEM	**Description**	Qty.	Price($)/unit	Total Cost
Standard compenents	Lifter rod	(Misumi)	1	20	20
	Lifter slide plate	(Misumi) - Oilfree	1	24.8	24.8
	Guide Rails for lifter slide core	(Misumi)- Oilfree	2	44	88
	Lifter Slide core (steel) - (Misumi)	(Misumi)	1	85.1	85.1
	Intermedium bushing for lifter rod	(Misumi)- Oilfree	1	27.5	27.5
		Total lifter manufacturing time =	380	Total	$245.40

Total lifter cost = Material+Manufacturing+STD parts   Total lifter cost ($)= $3,271.70

Milling Electorde (Graphite)

Mold info.	Undercut Depth	Ud(mm)	
	Surface Area of the undercut	As (cm^2)	120
	Projected area of the under cut (direction of the opening)	Up (cm^2)	100
	Undercut wall thickness	Ut (mm)	2
Values for steel 30-40 HRc (P-20 or similar)	Average removal rate (table speed)	Vf (mm/min)	600
	Depth of removal (axial depth 10% of D)	ap (mm)	0.05
	Effective diameter	De (mm)	0.30
	Cutter diameter	D (mm)	0.5
	Rotation speed of the cutter	RPM (rev/min)	20,000
	Chip load per tooth	Fz (mm/unit)	0.015
	Number of tooth or flutes on the cutter	Zn (units)	2
	Remove surface per minute	Ra (mm^2/min)	72
	Volumen removal rate	Qr (mm^3/min)	3.6
	Volumen to remove from core	Qc (mm^3)	240
	Irregular countour for machining	Qd (mm^3)	20
	Machining time (100 % efficiency)	Tc (h)	1.16
	Machining time (85 % efficiency)	**Tcf (h)**	**1.33**

$De = 2*(ap(D-ap))^{1/2}$

$Fz = (D*Hex)/De$

$Vf = RPM*Fz*Zn$

$Tcf (h) = (Qc/(Qr*60))*((0.5*Qd)/(Qr*60))$

$Ra = Vf*0.4*De$

$Qr = Ra*ap$

$Qc = As*Cd$

$Qd = (As-Ap)*Cd$

Total Cavity machining time (Tcf) = 1 hr

EDM

Mold info.	Volume to EDM from plastic part	Uv(mm^3)	120
	Average removal rate (Handbook of design,manufacturing, and automation)	Vr (mm^3/min)	180
	Machining time (100 % efficiency)	Tc (h)	0.67
	Machining time (85 % efficiency)	**Tedm (h)**	**0.77**

$Tc = Uv/Vr$

Total EDM time (Tedm) = 1 hr

Milling Lifter block (Steel)

Mold info.	Undercut Depth	Ud(mm)	
	Surface area of the undercut	As (cm^2)	120
	Projected area of the under cut (direction of the opening)	Up (cm^2)	100
	Undercut wall thickness	Ut (mm)	2
Values for steel 30-40 HRc (P-20 or similar)	Average removal rate (table speed)	Vf (mm/min)	600
	Depth of removal (axial depth 10% of D)	ap (mm)	0.05
	Effective diameter	De (mm)	0.30
	Cutter diameter	D (mm)	0.5
	Rotation speed of the cutter	RPM (rev/min)	20,000
	Chip load per tooth	Fz (mm/unit)	0.015
	Number of tooth or flutes on the cutter	Zn (units)	2
	Remove surface per minute	Ra (mm^2/min)	72
	Volumen removal rate	Qr (mm^3/min)	3.6
	Volumen to remove from core	Qc (mm^3)	240
	Irregular countour for machining	Qd (mm^3)	20
	Machining time (100 % efficiency)	Tc (h)	1.16
	Machining time (85 % efficiency)	**Tcf (h)**	**1.33**

$De = 2 \cdot (ap(D-ap))^{1/2}$ $\qquad$ $Ra = Vf \cdot 0.4 \cdot De$

$Fz = (D \cdot Hex)/De$ $\qquad$ $Qr = Ra \cdot ap$

$Vf = RPM \cdot Fz \cdot Zn$ $\qquad$ $Qc = As \cdot Cd$

$Tcf (h) = (Qc/(Qr \cdot 60)) \cdot ((0.5 \cdot Qd)/(Qr \cdot 60))$ $\qquad$ $Qd = (As-Ap) \cdot Cd$

Total Cavity machining time (Tcf) = **1** **hr**

Milling core block to make lifter housing (Steel)

Mold info.	Undercut Depth	Ud(mm)	
	Deptg of lifter block	As (mm^2)	120
	Projected area of the lifter (direction of the opening)	Up (mm^2)	100
	Undercut wall thickness	Ut (mm)	10
Values for steel 30-40 HRc (P-20 or similar)	Average removal rate (table speed)	Vf (mm/min)	600
	Depth of removal (axial depth 10% of D)	ap (mm)	0.05
	Effective diameter	De (mm)	0.30
	Cutter diameter	D (mm)	0.5
	Rotation speed of the cutter	RPM (rev/min)	20,000
	Chip load per tooth	Fz (mm/unit)	0.015
	Number of tooth or flutes on the cutter	Zn (units)	2
	Remove surface per minute	Ra (mm^2/min)	72
	Volumen removal rate	Qr (mm^3/min)	3.6
	Volumen to remove from core	Qc (mm^3)	1200
	Irregular countour for machining	Qd (mm^3)	20
	Machining time (100 % efficiency)	Tc (h)	5.60
	Machining time (85 % efficiency)	**Tcf (h)**	**6.44**

$De = 2 \cdot (ap(D-ap))^{1/2}$ $\qquad$ $Ra = Vf \cdot 0.4 \cdot De$

$Fz = (D \cdot Hex)/De$ $\qquad$ $Qr = Ra \cdot ap$

$Vf = RPM \cdot Fz \cdot Zn$ $\qquad$ $Qc = As \cdot Cd$

$Tcf (h) = (Qc/(Qr \cdot 60)) \cdot ((0.5 \cdot Qd)/(Qr \cdot 60))$ $\qquad$ $Qd = (As-Ap) \cdot Cd$

Total Cavity machining time (Tcf) = **6** **hr**

TOTAL SLIDE COST ($)

		Qty. of slides	1		

	ITEM	Description	Qty.	Price($)/unit	Total cost
MATERIAL	Graphite electrode for EDM	Dimensions (100mm, 200 mm, 30 mm)-Misumi	1	93.3	$93.30
	Slide block (H-13)	(Misumi) - Steel H-13 (4"x5"x6")	1	476	$476.00
		TOTAL MATERIAL COST			$569.30

	GROUP	Description	Time	$/hr	Total cost
MANUFATURING	Electrode	Set up cost	5	75	$375.00
		Machining cost (Milling)	1	75	$75.00
	EDM (rate 100 mm^3/min)	Set up cost	8	75	$600.00
		Electric discharge machining cost (EDM)	26	75	$1,950.00
	Machining slide block	Set up cost	7	75	$525.00
		Machining cost (Milling)	5	75	$375.00
	Machining core for slide housing	Set up cost	2	75	$150.00
		Machining core for lifter housing (Milling)	1	75	$75.00
	Machining cavity for cam shaft housing	Set up cost	2	75	$150.00
		Lifter rod housing (Drilling gun)	1	75	$75.00
		Total Slide manufacturing time =	58	Total	$4,350.00

	ITEM	Description	Qty.	Price($)/unit	Total Cost
Standard components	Angular pin	(Misumi)	1	73.3	$73.30
	Locking block	(Misumi)- Steel H-13 (4"x5"x3")	1	391	$391.00
	Guide rails	(Misumi) - Groove (10"x1"x1")	2	47.44	$94.88
	Sliding plate	(Misumi) - Groove (10"x5"x3/8")	1	112.4	$112.40
	Slide stopper	(Misumi) - (D=3/8", L=2")	1	10	$10.00
	Ball plunger	(Misumi) - (D=0.5", L=2")	1	7.2	$7.20
	Spring for slinde block	(Misumi) - (D=0.5", L=2.25")	2	3	$6.00
		Total slide manufacturing time =	58	Total	$694.78

Total slide cost = Material+Manufacturing+STD parts   **Total cost of the slide** $5,614.08

Milling Electorde (Graphite)

Mold info.	Surface Area of the undercut	As (cm^2)	120
	Projected area of the under cut (direction of the opening)	Up (cm^2)	100
	Undercut wall thickness	Ut (mm)	2
Values for steel 30-40 HRc (P-20 or similar)	Average removal rate (table speed)	Vf (mm/min)	600
	Depth of removal (axial depth 10% of D)	ap (mm)	0.05
	Effective diameter	De (mm)	0.30
	Cutter diameter	D (mm)	0.5
	Rotation speed of the cutter	RPM (rev/min)	20,000
	Chip load per tooth	Fz (mm/unit)	0.015
	Number of tooth or flutes on the cutter	Zn (units)	2
	Remove surface per minute	Ra (mm^2/min)	72
	Volumen removal rate	Qr (mm^3/min)	3.6
	Volumen to remove from core	Qc (mm^3)	240
	Irregular countour for machining	Qd (mm^3)	20
	Machining time (100 % efficiency)	Tc (h)	1.16
	Machining time (85 % efficiency)	**Tcf (h)**	**1.33**

$De = 2 * (ap(D-ap))^{1/2}$ $Ra = Vf*0.4*De$

$Fz = (D*Hex)/De$ $Qr = Ra*ap$

$Vf = RPM*Fz*Zn$ $Qc = As*Cd$

$Tcf (h) = (Qc/(Qr*60))*((0.5*Qd)/(Qr*60))$ $Qd = (As-Ap)*Cd$

Total electrode machining time (Tcf) =   1   hr

EDM (Slide)

Mold info.	Volume to EDM from plastic part	Uv(mm^3)	4000
	Average removal rate (Handbook of design,manufacturing, and automation)	Vr (mm^3/min)	180
	Machining time (100 % efficiency)	Tc (h)	22.22
	Machining time (85 % efficiency)	**Tedm (h)**	**25.56**

$Tc = Uv/Vr$

Total EDM time (Tedm) =   26   hr

Milling Slide block (Steel)

Mold info.	Surface Area of the undercut	As (cm^2)	220
	Projected area of the under cut (direction of the opening)	Up (cm^2)	200
	Undercut wall thickness	Ut (mm)	2
Values for steel 30-40 HRc (P-20 or similar)	Average removal rate (table speed)	Vf (mm/min)	600
	Depth of removal (axial depth 10% of D)	ap (mm)	0.05
	Effective diameter	De (mm)	0.30
	Cutter diameter	D (mm)	0.5
	Rotation speed of the cutter	RPM (rev/min)	20,000
	Chip load per tooth	Fz (mm/unit)	0.015
	Number of tooth or flutes on the cutter	Zn (units)	2
	Remove surface per minute	Ra (mm^2/min)	72
	Volumen removal rate	Qr (mm^3/min)	3.6
	Volumen to remove from core	Qc (mm^3)	440
	Irregular countour for machining	Qd (mm^3)	20
	Machining time (100 % efficiency)	Tc (h)	2.08
	Machining time (85 % efficiency)	Tcf (h)	2.40

De=2* (ap(D-ap))^ 1/2   Ra=Vf*0.4*De
Fz= (D*Hex)/De   Qr=Ra*ap
Vf= RPM*Fz*Zn   Qc=As*Cd
Tcf (h)=(Qc/(Qr*60))*((0.5*Qd)/(Qr*60))   Qd=(As-Ap)*Cd

Total slide machining time (Tcf) =   2   hr

Milling core block to make slide housing (Steel)

Mold info.	Surface Area of the undercut	As (cm^2)	220
	Projected area of the slide (direction of the opening)	Up (cm^2)	200
	Undercut wall thickness	Ut (mm)	2
Values for steel 30-40 HRc (P-20 or similar)	Average removal rate (table speed)	Vf (mm/min)	600
	Depth of removal (axial depth 10% of D)	ap (mm)	0.05
	Effective diameter	De (mm)	0.30
	Cutter diameter	D (mm)	0.5
	Rotation speed of the cutter	RPM (rev/min)	20,000
	Chip load per tooth	Fz (mm/unit)	0.015
	Number of tooth or flutes on the cutter	Zn (units)	2
	Remove surface per minute	Ra (mm^2/min)	72
	Volumen removal rate	Qr (mm^3/min)	3.6
	Volumen to remove from core	Qc (mm^3)	440
	Irregular countour for machining	Qd (mm^3)	20
	Machining time (100 % efficiency)	Tc (h)	2.08
	Machining time (85 % efficiency)	Tcf (h)	2.40

De=2* (ap(D-ap))^ 1/2   Ra=Vf*0.4*De
Fz= (D*Hex)/De   Qr=Ra*ap
Vf= RPM*Fz*Zn   Qc=As*Cd
Tcf (h)=(Qc/(Qr*60))*((0.5*Qd)/(Qr*60))   Qd=(As-Ap)*Cd

Total Core block machining time (Tcf) =   2   hr

Group 2

	Formula	Variables	Value	Tc (H)	Mch ($/H)	Cch ($)
Cooling System	Tc=n*0.22*L	L (in)	9	1.98	75	$148.50
Leader bushing housing	Ti=2.36*D	D (in)	3	7.08	75	$531.00
Ejector pin housing	Tj=0.036*n*(L-S)	n(unit)	14	15.12	75	$1,134.00
		L (in)	50			
		S(in)	20			
Sleeve pin housing	Tj=0.028*n*(L-S)	n(unit)	14	11.76	75	$882.00
		L (in)	50			
		S(in)	20			
Return pin housing	Tj=0.055*n*(L-S)	n(unit)	4	6.6	75	$495.00
		L (in)	50			
		S(in)	20			
Sprue puller housing	Tsp=0.036*n*(L-S)	n(unit)	1	1.08	75	$81.00
		L (in)	50			
		S(in)	20			

Group 3

	Formula	Variables	Value	Tc (H)	Mch ($/H)	Cch ($)
Mold Design	Small mold=64	hr	64	168	75	$12,600.00
	Medium mold=120	hr	120			
	Big mold=168	hr	168			
Programing	Tp=(Tc1+tc2+Ts+Ti)*Kp	Tc1	43	139.82	75	$10,486.28
		Tc2	43			
		Ts	380			
		TI	0			
		Kp	0.3			
Benching	Small mold=16	hr	40	70	75	$5,250.00
	Medium mold=24					
	Big mold=40					
	Time per action	hr/units	3			
	Qty. of slides	Units	0			
	Qty. of lifters	Units	10			
Spoting	Small mold=4	hr	8	30	75	$2,250.00
	Medium mold=6					
	Big mold=8					
	Time per action	hr/units	2			
	Qty. of slides	Units	1			
	Qty. of lifters	Units	10			
Polishing	Polishing rate = 120	mm^2/h	120	30.833	75	$2,312.50
	Surface Area of cavity (As)	mm^2	3700			
Assembly	Small mold=4	hr	10	10	75	$750.00
	Medium mold=6					
	Big mold=10					
Mold Trial	Small mold=600	$/trial	1000	N/A	N/A	$3,000.00
	Medium mold=650					
	Big mold=1,000					
	Number of mold trials	trial	3			
Transportation	Distance	Miles	****	N/A	N/A	$1,200.00
	Mold Weight	Lb	****			
	Truck rate	$	1200			
Proceesing fees	Total mold cost	$		N/A	N/A	TBD
	(%) of Processing fees (10%)	%	0.1			

FINAL COST OF THE PLASTIC INJECTION MOLD

VARIABLE	DESCRIPTION	COST ($)
A	Total cost for standard componets ($)	A
B	Total cost for part manufactured per the mold maker ($)	B
C	Processing fees (10% of total cost) ($)	C = (A + B) x (0.10)
D	Tolerance of deviation in the cost estimation (10% of total cost) ($)	D = (A + B + C) x (0.10)
E	Over head and unexpected activities (10%)	E = (A + B + C + D) x (0.10)
F	Profit (15% of total cost)	F = (A + B + C + D + E) x (0.15)
	FINAL COST OF THE PLASTIC INJECTION MOLD =	(A+B+C+D+E+F)

FINAL COST OF THE PLASTIC INJECTION MOLD

VARIABLE	DESCRIPTION	COST ($)
A	Total cost for standard componets ($)	$97,115.88
B	Total cost for part manufactured per the mold maker ($)	$43,374.55
C	Processing fees (10% of total cost) ($)	$14,049.04
D	Tolerance of deviation in the cost estimation (10% of total cost) ($)	$15,453.95
E	Over head and unexpected activities (10%) ($)	$16,999.34
F	Profit (15% of total cost) ($)	$28,048.92
	FINAL COST OF THE PLASTIC INJECTION MOLD =	$215,041.68

Glossary

CAD: Computer aided design.

CAE: Computer aided engineering.

CAM: Computer aided machining.

Capacity: The upper limit or ceiling on the load that an operating unit can handle.

Cavity: A hollow space within the mold where one side of the plastic part(s) is shaped.

CNC: Computer numerical control.

Concurrent engineering: Bringing engineering design and manufacturing personnel together during in the design phase.

Cooling channels: Machined channels throughout the mold to circulate a fluid in order to remove the heat from the part(s).

Core: The part of the mold that fills the hollow space of the cavity shaping the other half of the plastic part(s).

Cycle time (molding): The required time to complete the sequence for producing a plastic part.

Ejector pin: These pins remove the plastic part from the mold.

Ejection system: The components of the mold responsible for ejecting the plastic parts from the mold.

Finishing: The level of roughness of the mold and/or the part.

Gate: The area through which the melted plastic enters into the cavity and core section of the mold where the plastic part is shaped.

Injection Molding: The process by which objects are formed when a plastic material is fed through a nozzle into a mold where it is held until removed in a solid state, duplicating the cavity of the mold.

Lead-time: The time it takes to design and build a mold; often referred to as the estimated time of the completion of a mold.

Master production schedule: Statement of what will be made, how many units will be made, and when they will be made.

Net Investment: The project's initial net cash outlay, that is, the outlay at time period cero.

Planning: Determining what needs to be done, by whom, and by when, in order to fulfill one's assigned responsibility.

Plastic injection mold: The cavity, core, and base components that comprise the tool in which a plastic part is formed or molded.

Productivity: A measure of the effective use of resources, usually expressed as the ratio of output to input.

Project management: The process of planning, organizing, directing, and controlling of company resources to complete specific goals and objectives.

Project risk: Measure of the probability and consequence of not achieving a defined project goal.

Quality: The ability of a product or service to consistently meet or exceed customer expectations.

Runner system: The channel through which the melted plastic flows to the cavity.

Sprue: The opening in an injection mold through which the melted plastic is fed from the nozzle of the injection machine to the runner.

Strategic planning for project management: The development of a standard methodology for project management, which can be used over and over again, and which will produce a high likelihood of achieving the project's objectives.

BIBLIOGRAPHY

Gido, J., Clements, J. (2nd ed.). (2003).

Successful Project Management. Mason, OH:

South-Western.

Goldsberry, C. (1st ed.). (2000). Purchasing

Injection Molds: A Buyer's Guide. Denver, CO:

Injection Molding Magazine.

Kerzener, H., PhD. (7th ed.). (2001).

Project Management: A System Approach to

Planning, Scheduling, and Controlling. New York,

NY: John Wiley & Son, Inc.

Menges, G., Michaeli, W., & Mohren, P.

(3rd ed.). (2001). How to Make Injection Molds.

Cincinnati, OH: Hansner Gardner Publications,

Inc.

Moyer, C., McGuigan, J., & Kretlow, W. (8th ed.). (2001). <u>Contemporary Financial Management</u>. Cincinnati, OH: SouthWestern College Publishing.

Rosato, D., Rosato, D. (2nd ed.). (1999). <u>Injection Molding Handbook</u>. Norwell, MA: K. Kluwer Academic Publishes.

Shillinglaw, Gordon; McGaharan. (9th ed.). (1993). <u>Accounting: A Management Approach</u>. Home Wood, IL: Irwin.

Stevenson, W. (7th ed.). (2002). <u>Operation Management</u>. New York, NY: McGraw-Hill/Irwing